AF566877

Queueing Networks with Blocking

QUEUEING NETWORKS WITH BLOCKING

Exact and Approximate Solutions

Harry G. Perros

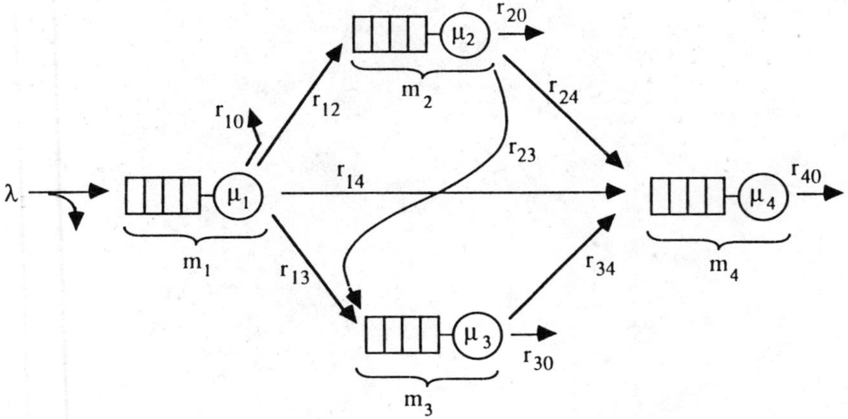

New York Oxford
OXFORD UNIVERSITY PRESS
1994

Oxford University Press

Oxford New York Toronto
Delhi Bombay Calcutta Madras Karachi
Kuala Lumpur Singapore Hong Kong Tokyo
Nairobi Dar es Salaam Cape Town
Melbourne Auckland Madrid

and associated companies in
Berlin Ibadan

Published by Oxford University Press, Inc.,
200 Madison Avenue, New York, New York 10016

Library of Congress Cataloging-in-Publication Data
Perros, Harry G.
Queueing networks with blocking: exact and approximate solutions / Harry G. Perros.
p. cm. Includes bibliographical references and index.
ISBN 0-19-508580-9
1. Queueing theory. I. Title.
T57.9.P38 1994 004.2'4'0151982—dc20 94-8089

9 8 7 6 5 4 3 2 1
Printed in the United States of America
on acid-free paper

To

Helen, Nicholas, and Mikey
for all they have given me

and

to the memory of my mother

Forward

Queueing networks with blocking provide a useful modelling tool for evaluating the performance of systems such as computers, communications networks, and production lines. Open and closed queueing networks with blocking have been extensively analyzed by researchers from various research communities, such as Operations Research, Computer and Communication Performance Modelling, and Industrial Engineering. In view of this, related results are scattered throughout various journals and conference proceedings. In this book I have attempted to present some of these results in a systematic fashion. I hope that this book will be of use to researchers and graduate students who would like to know more about this area. Also, it is my sincere hope that it will be of help to practitioners involved with the performance evaluation of computers, communication networks, and production systems.

It was often difficult to choose which results to include in a Chapter. In order to make up for my subjective choice, I have included at the end of each Chapter a short description of other relevant results and a bibliography. I have assumed that the reader has knowledge of queueing theory, equivalent to about one semester of queueing theory. The outline of the book is as follows. In Chapter 1, some of the basic concepts regarding the various blocking mechanisms are described, followed by a review of the phase-type distribution which is used extensively throughout the book. Chapter 2 contains a short review of equation solving techniques used to analyze continuous time Markov processes with discrete states. These techniques are used in some of the approximation algorithms described in the book. Chapters 3 to 5 deal with the analysis of open queueing networks with blocking. Specifically, in Chapter 3, the simplest open queueing network with blocking consisting of two nodes linked in tandem is discused. Various exact solutions for this simple queueing network under different assumptions regarding the service times are examined. Chapter 4 deals with the analysis of open tandem queueing networks with blocking. We present a variety of decomposition algorithms and some exact results for constant service times. Finally, in

Chapter 5, two different decomposition algorithms for the analysis of arbitrarily linked open queueing networks with blocking are presented. Chapters 6 and 7 deal with closed queueing networks with blocking. In Chapter 6, exact product-form solutions, based on the notion of time reversibility, are presented. In Chapter 7, various approximation procedures for the analysis of closed queueing networks with blocking are described. The applicability of queueing networks with blocking to modelling real-life systems, is demonstrated in Chapter 8 which contains models for a production system, a disk I/O system, and a multi-stage Banyan ATM switch. Following Chapter 8, a bibliography of exact and approximate results for open and closed queueing networks with blocking is given. It contains most of the references up until the summer of 1993. Finally, a table of symbols is given at the end of the book.

I would like to thank my good friend and co-author, Tayfur Altiok. Originally, we started out to write this book together. Unfortunately, our collaboration on this book was not to last very long because of our different priorities. I still recall fondly the heated discussion we had one summer day in 1989 in a small café in *Quartier Latin* over the layout of the book! He suggested to me to include a Chapter on the numerical techniques of Markov Chains, and he wrote an early version of section 2.4. He also wrote the general solution of the matrix-recursive procedure and the matrix-geometric procedure in section 3.4. Finally, he gave me constructive criticisms on Chapters 1 to 4. I am indebted to Simoneta Balsamo for reviewing the entire manuscript. I am also indebted to the following referees for reviewing parts of the manuscript: Yves Dallery, Yannick Frein, Dimitri Kouvatsos, Vittoria De Nitto Personè, Billy Stewart, Hideaki Yamashita, and my students Khaled El-Sayed and Young Rhee. Also, I would like to thank Raif Onvural for his continuous encouragement! Last but not least, I would like to thank Margaret Hudacko for expertly editing the camera-ready version of this book. Finally, it would be unjust if I did not mention Bilbo, our four-legged member of our familly, who never failed to fall asleep by my feet while I was labouring on this book!

Contents

Queueing Networks with Blocking

CHAPTER 1

Basic Concepts

1.1. Introduction

Queueing networks have been used successfully to model real-life systems. A queueing network is a set of interconnected nodes. Each node consists of a queue, where customers wait for service, and one or more servers. Queueing networks have been studied extensively under a multiplicity of assumptions regarding service and inter-arrival distributions, multiple classes, service disciplines, priorities, etc. A common assumption made when analyzing queueing networks is that the capacity of each node is infinite. That is, a node can accommodate any number of customers waiting to receive service. However, in real-life systems frequently the storage space in front of a server is finite. This observation has given rise to the study of queueing networks with finite capacity nodes. Due to the limitations imposed on the capacity of these nodes, the flow of customers through one node may be momentarily stopped if a destination node has reached its capacity. This is known as blocking, and in view of this, a queueing network with finite capacity nodes is referred to as a *queueing network with blocking*. When a customer in a node gets blocked, the customer is unable to proceed to its destination node. The customer is not lost, however. Rather, it is delayed for a period of time. We note that this type of blocking is not related to the notion of blocking in teletraffic, where an arriving customer to a node is blocked, i.e. lost, if the node is full.

Queueing networks with blocking have proved useful in modelling production systems, computer systems, distributed systems, and telecommunication systems. Below, we briefly describe a few real-life systems which have been successfully modelled by queueing networks with blocking.

A production line: A production line consists of a set of machines, P_1 to P_n, arranged in series as shown in figure 1.1. In front of each machine P_i there is a storage space ST_i. Work-pieces visit each machine in sequence before they finally leave the production line. In particular, they move from machine P_i to storage space ST_{i+1} to machine P_{i+1}. Each storage space can only hold a finite number of work-pieces. Machine P_i gets blocked, i.e. it cannot process any work-pieces in its storage space ST_i , when the downstream storage space ST_{i+1} becomes full. The machine remains blocked until some space becomes available in the downstream storage space. Such a production system is typically modelled by a tandem configuration of finite capacity nodes. Various solution techniques for analyzing tandem configuration of finite capacity nodes are given in Chapter 4.

A flexible manufacturing system: Consider a flexible manufacturing system consisting of a network of automated work stations linked by a material handling device (MHS), as shown in figure 1.2. Each work station consists of a finite capacity storage space and one or more machines. There is also a finite capacity central storage. The material handling device transports work-pieces between work stations, as well as between work stations and the central storage. A constant number of work-pieces circulate in the system. There is always at least one work-piece waiting to enter the flexible manufacturing system. Thus, when a finished work-piece leaves the system, another takes its place. Upon completion of the processing of a work piece, the work-piece is forwarded to the next work station. If the storage space of the next work station is full, the work-piece is forwarded to the central storage. If the central storage is also full, the work-piece stays in front of the machine, and the machine gets blocked. The machine will remain blocked until the work-piece departs from its station. Such flexible manufacturing systems have been modelled as closed queueing networks with blocking. Exact and approximate solutions of closed queueing networks with blocking are given in Chapters 6 and 7 respectively.

Figure 1.1: A production line.

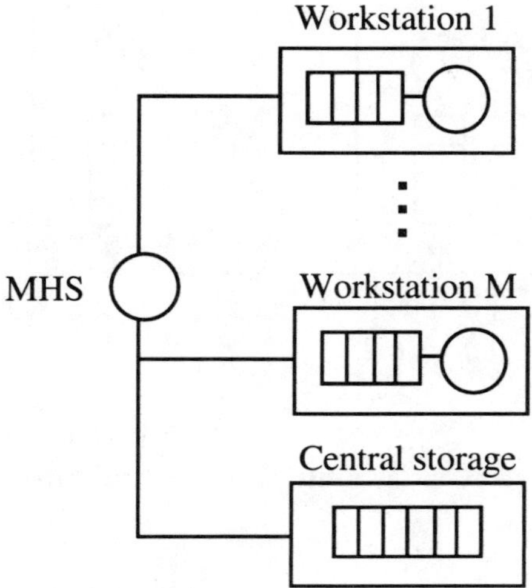

Figure 1.2: A flexible manufacturing system.

A multiprocessor architecture: Consider a multiprocessor system consisting of M processors and N memory modules. The processors and the memory modules are connected together by a bus, as shown in figure 1.3. Each memory module has a finite capacity input buffer for queueing memory access requests. Each processor has its own local memory, but it can also access the N memory modules. During the time the processor is accessing a memory module, the processor is blocked and it cannot do any processing. In order for processor i to access memory module j, the processor must first get hold of the bus. If the bus is not busy, processor i transfers its memory access request to memory module j over the bus. If the bus is busy then processor i keeps retrying until it get hold of the bus. When the memory access request arrives at memory module j, the request is queued. If the queue of memory module j is full the request is not accepted and processor i has to retransmit it at a later time. When a memory module completes a request, it attempts to get hold of the bus and it then transmits its answer back to the appropriate processor. During the time the memory module waits for the bus, the memory module cannot execute any other requests. Depending upon traffic assumptions, this multiprocessor architecture can be analyzed using either a closed or an open queueing network with blocking.

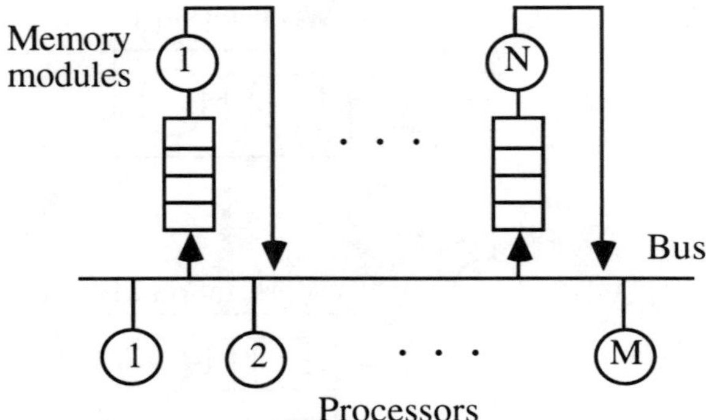

Figure 1.3: A multiprocessor architecture.

A virtual circuit: In a packet switched network, packets may be transferred between two users using a virtual circuit or datagrams. When a virtual circuit is used, a path is established between these two users over a number of nodes of the packet switched network. Each node allocates a finite capacity buffer for storing incoming packets. When the buffer of a node is full, the node does not accept any more packets. During this time its upstream node gets blocked, and it cannot transmit packets to this node. Virtual circuits have been modelled as open queueing networks with blocking. Various techniques for analyzing open queueing networks with blocking are presented in Chapters 4 and 5.

In general, queueing networks with blocking are difficult to analyze. Under certain assumptions they can be shown to have an elegant product-form solution. However, in most cases closed-form solutions are not available. In view of this, it is a common practice to analyze queueing networks with blocking approximately using decomposition techniques. That is, the queueing network is broken up into sub-systems, and each sub-system is analyzed in isolation. The results obtained from the sub-systems are combined together through the means of an iterative scheme. An alternative solution technique is to formulate a queueing network with blocking as a Markov process. The steady-state probability vector is then obtained using numerical techniques for systems of linear equations.This approach gives exact numerical results, but it is limited to queueing networks whose underlying Markov process has a manageable state space. Queueing networks with blocking can also be analyzed by simulation. This method is considerably more cpu intensive than any of the above mentioned approaches. It is, however, the most flexible modelling technique since one can practically simulate a queueing system under realistic assumptions. In this book, we examine mainly exact and approximation solutions for the analysis of queueing networks with blocking. We also examine to some extent Markov chain numerical solution techniques.

In this Chapter, we discuss some basic concepts that will be used throughout the book. In particular, in section 1.2, we introduce three different types of blocking mechanisms. Equivalencies between these three blocking mechanisms are discussed in section 1.3. In section 1.4, we review some of the properties of the phase-type distribution and in particular of the Coxian distribution, and in section 1.5, we discuss how a general distribution can be approximated by a Coxian distribution. Finally, relevant references are given in section 1.6.

1.2. Blocking mechanisms

Before we proceed, we note that a finite capacity node will be indicated by the familiar symbol given in figure 1.4a. The server is represented by a circle and the finite capacity queue is represented by a rectangle consisting of a finite number of positions. It is assumed that there is a position in front of the server that is occupied by the customer currently in service. Thus, the maximum number of customers, m, that can be accommodated in this finite capacity node is equal to the maximum number of customers m-1 that can be accommodated in its queue plus the one in service. An infinite capacity node, shown in figure 1.4b, will be indicated in the same way, only the rectangle is open-ended at the left-hand-side, thus indicating an infinite number of positions.

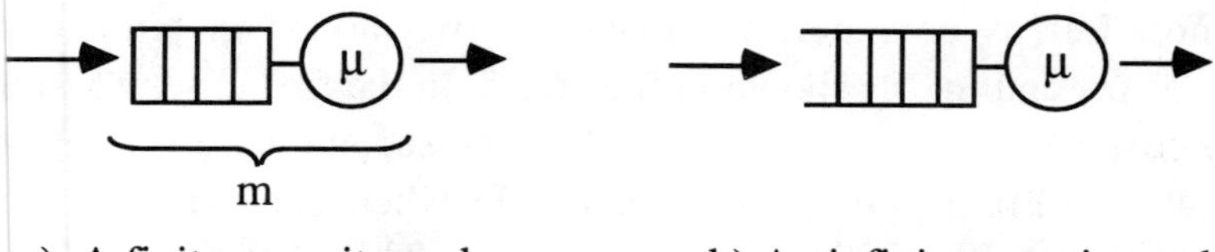

a) A finite capacity node

b) An infinite capacity node

Figure 1.4: A single node.

A blocking mechanism is a set of rules that dictates the blocking and unblocking of a server. Various blocking mechanisms have been considered in the literature so far. These mechanisms arose out of different studies of real-life systems. They are distinct types of models for blocking, a fact that may be easily missed by a reader unfamiliar with the subject. The following are the most commonly used blocking mechanisms.

Blocking-after-service: A customer upon service completion at node i attempts to join destination node j. If node j at that moment is full, the customer is forced to wait in node i, in front of server i, until it enters destination node j. Server i becomes blocked for this period of time and it cannot serve any other customer waiting in the node. This blocking mechanism has been used to model systems such as disk I/O configurations and production lines. In the literature, this blocking mechanism has been referred to by a variety of different names such as type 1 blocking, transfer blocking, production blocking, and non-immediate blocking.

Blocking-before-service: A customer in node i declares its destination (say node j) prior to starting service. If node j is full, the server of node i becomes blocked, i.e. it cannot serve other customers. When a departure occurs from destination node j, the server of node i becomes unblocked and the customer begins receiving service. This blocking mechanism has been used to model production lines and telecommunication systems, such as virtual circuits. In the literature, it has been referred to by names such as type 2 blocking, communication blocking, immediate blocking, and service blocking.

Depending upon whether the blocked customer is allowed to occupy the position in front of the server or not, one can distinguish two cases. In the first case, the position in front of the server can be occupied by the blocked customer when the server is blocked, whereas in the second case it cannot. Let us consider, for instance, two nodes in tandem with capacities m_1 and m_2, as shown in figure 1.5. In the first case, when there are m_2 customers in the second node (i.e. the second node is full), there can be at most m_1 customers in the first node, whereas in the second case, there can be at most m_1-1 customers in it. This distinction may be meaningful in certain real-life systems. We note that the first case is the most common one. In view of this, unless otherwise specified, in this

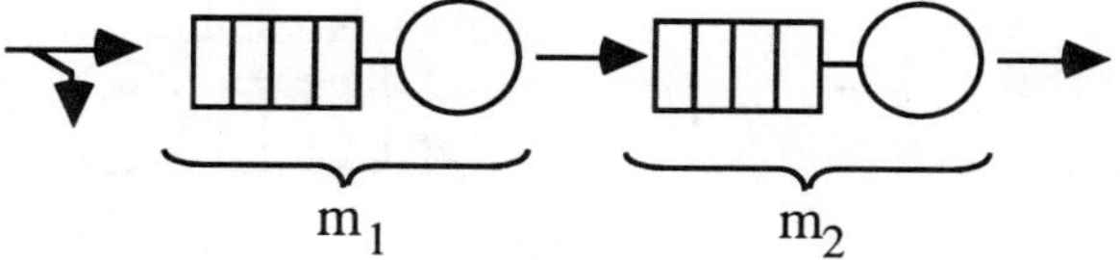

Figure 1.5: Two nodes in tandem.

book we shall assume that in the blocking-before-service mechanism, the blocked customer is allowed to occupy the position in front of the server.

Repetitive-service: A customer upon service completion at node i attempts to join node j. If node j is full, the customer immediately receives another service at node i. This is repeated until the customer completes a service at node i at a moment when node j is not full. In the literature, this type of blocking has been referred to by names such as type 3 blocking, and rejection blocking.

In the above description, it was assumed that the destination node is fixed. In general, one can distinguish two cases: fixed destination and random destination. In the first case, once the customer's destination is determined it cannot be altered. That is, each time the customer completes a service it attempts to enter the same destination node. In the second case, each time the customer completes a service, a destination node is chosen independently of the destination node chosen the previous time. The fixed destination case has been used to model telecommunication systems, and the random destination case has been used to model flexible manufacturing systems.

To understand how these blocking mechanisms work, we consider a queueing network with blocking consisting of a node (node 0) linked to M parallel nodes (numbered 1 to M) as shown in figure 1.6. We shall refer to this configuration as *one-to-many* configuration. Customers arrive from outside and join node 0. A customer, upon service completion at node 0 joins one of the M downstream nodes with probability r_{0i}, i=1,2,...,M, and after service completion it departs from the network. In this configuration, the server at node 0 may get blocked by any of the M downstream nodes. In the case of blocking-after-service, a customer at node 0 will choose a destination node j, j=1,2,...,M, upon completion of its service. If node j is full, the customer will remain at node 0 until the moment when a departure occurs from node j. Then, the customer will

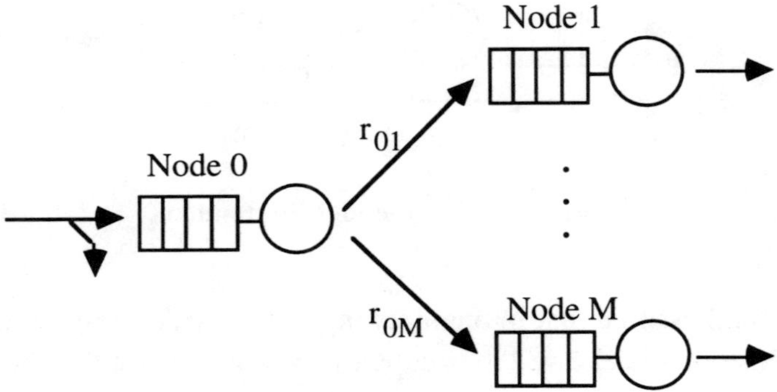

Figure 1.6: One-to-many configuration.

join node j and the server at node 0 will become unblocked. In the case of blocking-before-service, a customer will first choose a destination node j, j=1,2,...,M, and if this node is not full it will begin service. If node j is full, the server at node 0 will become blocked. It will remain blocked until a departure occurs from node j. If the blocking mechanism is repetitive-service, then the customer in node 0 will choose a destination node j, j=1,2,...,M, upon completion of its service. If node j is full, the customer will immediately start a new service at node 0. In the fixed destination case, the customer will repeatedly receive service at node 0 until it succeeds to enter node j. In the random destination case, however, the customer will repeatedly receive service at node 0, but each time it completes a service it will choose a new destination node independently of the destination node chosen last time.

Finally, we need to clarify further the mechanisms blocking-after-service and blocking-before-service when there are several nodes which have the same destination node. For this, we consider a queueing network consisting of M nodes in parallel (numbered 1 to M) all linked to the same node (node 0) as shown in figure 1.7. We shall refer to this configuration as *many-to-one* configuration. There are M independent arrival streams, one for each of the M nodes. A customer at node j, j=1,2,...,M, upon completion of its service joins node 0 with probability 1, and after receiving service, it departs from the queueing network. In this configuration, nodes 1 to M can get blocked. In the case of blocking-after-service, when a customer completes its service at the ith node, i=1,2,...,M, at a time that node 0 is full, the customer becomes blocked. At that instant, there may be other customers (from the other M-1 nodes) also blocked. Blocked customers enter node 0 on a *first-blocked-first-enter* basis. Unblocking of the node i's server occurs when the blocked customer finally enters node 0. Now, let us consider the case of blocking-before-service. In this case, when node 0 becomes full, all upstream services currently in progress are interrupted. These blocked services are resumed when a departure occurs from node 0. That is, a customer at node i, i=1,2,...,M, may become blocked at any time during its service if node 0 becomes full (due to arrivals from the other M-1 nodes). During a blocking period, a customer that arrives at an idle node i, i=1,2,...,M, cannot get served. The case of blocking-before-service where the blocked customer is not allowed to occupy the position in front of the server cannot be defined in a many-to-one configuration. This is because a node, say node i, maybe full when node 0 becomes full due to an arrival from another node. At that moment, all upstream nodes including node i will become blocked. However, the blocking customer at node i will still occupy the position in front of the server, which is inconsistent with the definition of this blocking mechanism!

An interesting situation occurs in the case of blocking-after-service, when a blocked node is itself blocking another node. We demonstrate this case using a tandem configuration. This configuration consists of M, M≥2, finite capacity nodes arranged in series as shown in figure 1.8. Customers arrive from outside and join node 1. Each customer visits nodes 1 to M in sequence, and eventually it departs from node M.

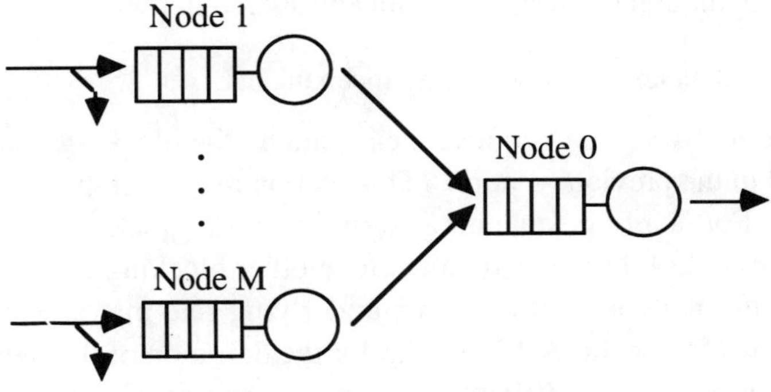

Figure 1.7: Many-to-one configuration.

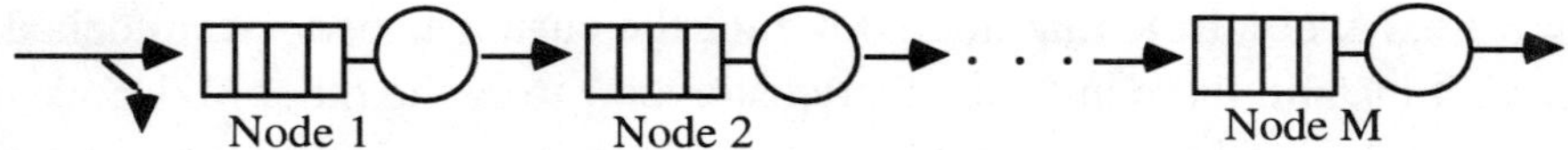

Figure 1.8: A tandem configuration.

Blocking-after-service is assumed. Let us consider the case where node i is blocked by node i+1, which in turn is blocked by node i+2. When a departure occurs from node i+2, the blocked customer from node i+1 will move into node i+2. This will free a space in node i+1, which will permit the blocked customer from node i to move into node i+1. We note that the two blocked customers will move into the next node at the same time. Thus, a departure from node i+2 will cause both nodes i and i+1 to get unblocked simultaneously. This situation of consecutive blocked nodes may involve several nodes at a time. For instance, nodes i to i+k, k≥1, may all be blocked at the same time. These nodes will all become unblocked simultaneously at the moment when a departure occurs from node i+k+1!

In the above three blocking mechanisms, a server becomes unblocked when the number of customers in the destination node drops below its maximum capacity. Extensions of these mechanisms have been considered where unblocking occurs when the number of customers in the destination node drops below a pre-defined level, not necessarily equal to its maximum capacity. Other blocking mechanisms associated with queueing networks with blocking which have a product-form solution will be examined in Chapter 6. Finally, we note that queueing networks with sub-network population constraints are not considered in this book.

1.3. Equivalencies among blocking mechanisms

In this section, we obtain equivalencies among the blocking mechanisms described in the previous section. (This section may be skipped in the first reading.) For a given queueing network with blocking, a blocking mechanism i is said to be equivalent to another blocking mechanism j, if the queueing network has the same underlying rate matrix under either mechanism. (See sections 2.2 and 2.3 for the defintion of the rate matrix). In some cases, i is equivalent to j if the capacities of the nodes of the queueing network are reduced by one. In this section, for presentation purposes, we will refer to the blocking mechanisms by the abbreviations given in Table 1.1. The equivalencies will be obtained for some specific configurations of open and closed queueing networks with blocking.

Open queueing networks with blocking

We will consider three basic configurations, namely, the one-to-many configuration shown in figure 1.6, the many-to-one configuration shown in figure 1.7, and the tandem configuration shown in figure 1.8. For each of these three configurations, we will assume exponentially distributed service times and Poisson arrivals. Most of the equivalencies presented below are intuitively obvious. For instance, in a tandem configuration each node has only one destination. Therefore, it is does not matter if we assume a repetitive-service blocking mechanism with a random destination or with a fixed destination. The same holds for the many-to-one configuration. The following equivalence for tandem configurations however, is not as obvious. This equivalence is obtained for a three-node tandem configuration. The proof can be easily extended to any number of nodes.

Theorem: Consider a tandem configuration (call it A) with bbs-sno and capacities m_1, m_2, and m_3. Let us consider now the same configuration with bas and capacities m_1, m_2-1 and m_3-1 (call it B). Then, configurations A and B have the same rate matrix.

Table 1.1: Abbreviations of the names of the blocking mechanisms.

Blocking mechanism	Abbreviation
Blocking-after-service	bas
Blocking-before-service	
position in front of server can be occupied	bbs-so
position in front of server cannot be occupied	bbs-sno
Repetitive-service	
fixed destination	rs-fd
random destination	rs-rd

Proof: We will obtain the equivalence between these two blocking mechanisms by comparing their associated rate matrices. In view of this, we start by describing the state space of the tandem configuration under bas and bbs-sno. The state of the tandem configuration under bas can be described by (n_1,n_2,n_3), where n_i is the number of customers in node i, i=1,2,3. We have $n_1=0,1,\ldots,m_1$, $n_2=0,1,\ldots,m_2$, and $n_3=0,1,\ldots,m_3$, where $n_2=m_2$ denotes that node 1 is blocked by node 2, and $n_3=m_3$ denotes that node 2 is blocked by node 3. In the case of bbs-sno, the tandem configuration is described by a similar state vector (n_1,n_2,n_3), where n_i is the number of customers in node i, i=1,2,3, and they take the values $n_1=0,1,\ldots,m_1$, $n_2=0,1,\ldots,m_2$, and $n_3=0,1,\ldots,m_3$. In addition, we have the constraints $n_1 \le m_1-1$ if $n_2=m_2$ and $n_2 \le m_2-1$ if $n_3=m_3$.

Let us now apply the following transformations to the states of the tandem configuration A.

a) $(n_1,n_2,n_3) \rightarrow (n_1,n_2,n_3)$ for the states (n_1,n_2,n_3), $n_1=0,1,\ldots,m_1$, $n_2=0,1,\ldots,m_2$, $n_3=0,1,\ldots,m_3$, in which no node is blocked.
b) $(n_1,n_2,m_3) \rightarrow (n_1,n_2+1,m_3)$ for the states (n_1,n_2,m_3), $n_1=0,1,\ldots,m_1$, $n_2=0,1,\ldots,m_2-1$, in which only node 2 is blocked.
c) $(n_1,m_2,n_3) \rightarrow (n_1+1,m_2,n_3)$ for the states (n_1,m_2,n_3), $n_1=0,1,\ldots,m_1-1$, $n_3=0,1,\ldots,m_3-1$, in which only node 1 is blocked.
d) $(n_1,m_2-1,m_3) \rightarrow (n_1+1,m_2,m_3)$ for the states (n_1,m_2-1,m_3), $n_1=0,1,\ldots m_1-1$, in which both nodes are blocked.

We note that the transformed states are those associated with configuration A. The above transformation is one-to-one and onto between the set of states associated with configuration A and B. Also, transition rates to and out of a state in A and its associated state in B are the same. Hence, the two configurations have the same rate matrix. This result holds, irrespective of whether the capacity of node 1 is finite or infinite.□

We now proceed to obtain some other equivalencies using simple intuitive arguments. For each of the three configurations considered here bbs-so and rs-fd are equivalent. For each configuration, this can be shown to be true by examining the rate matrix associated with bbs-so and rs-fd. It can be easily verified that the only additional transitions in the rate matrix of rs-fd which are not in the rate matrix of bbs-so are feedbacks (from a state back to the same state) due to blocked departures. These transitions cancel out in the rate matrix, and therefore, both blocking mechanisms have the same rate matrix.

In the one-to-many configuration, bas is not equivalent to any of the other blocking mechanisms. This can be shown by simply observing that we cannot obtain an equivalence between bas and bbs-sno and bas and bbs-so. This is because in bbs-sno and bbs-so, a customer in node 0 chooses its destination node before it starts its service. In view of this, it is not possible to find an equivalence even by adjusting the buffer sizes of nodes 1 to M. Similarly, bas is not equivalent to rs-rd, since under bas a customer's destination is not changed. Finally, bas is not equivalent to rs-fd since, as it was shown above, rs-fd is equivalent to bbs-so and bbs-so is not equivalent to bas.

In the one-to-many configuration, rs-rd is not equivalent to any other type. This is true by definition. Since the destination of a customer may be changed at each service completion, the rate matrix for rs-rd is different to that of any other blocking type.

In the one-to-many configuration, if node 0 has an infinite capacity, then bbs-sno and bbs-so are equivalent. The same holds for the many-to-one configuration if nodes 1 to M have infinite capacity. The only difference between the two blocking mechanisms is that in bbs-sno the customer is not allowed to occupy the position in front of the server. This has no effect since the queue has an infinite capacity. However, if node 0 has a finite capacity, then bbs-sno and bbs-so are not equivalent. The same result holds for the many-to-one configuration if nodes 1 to M have a finite capacity.

In the many-to-one configuration, bas is not equivalent to any other blocking mechanism. The state vector for this configuration under bas contains the order in which the servers of nodes 1 to M get blocked. This does not exist in the state vector of any of the remaining blocking mechanisms. Hence, it is not possible to obtain the same rate matrix under any of the remaining blocking mechanisms.

Below we summarize the above equivalencies.

Tandem configurations:

a) bbs-sno with node capacities m_1, m_2,...,m_M, is equivalent to bas with node capacities m_1, m_2-1,...,m_M-1. (m_1 may be finite or infinite.)

b) bbs-so, rs-fd and rs-rd are equivalent.

Two-node tandem configurations:

a) bbs-sno (or bbs-so, or rs-fd, or rs-rd) with node capacities $m_1=\infty$ and m_2 is equivalent to bas with node capacities $m_1=\infty$ and m_2-1.
b) bbs-sno, bbs-so, rs-fd and rs-rd are equivalent if $m_1=\infty$.

One-to-many configurations:

a) bbs-so and rs-fd are equivalent if node 0 has a finite capacity.
b) bbs-so, bbs-sno and rs-fd are equivalent if node 0 has infinite capacity.

Many-to-one configurations:

a) bbs-so, rs-fd and rs-rd are equivalent if nodes 1 to M have a finite capacity.
b) bbs-so, bbs-sno, rs-fd and rs-rd are equivalent if nodes 1 to M have infinite capacity.

Closed queueing networks with blocking

The equivalencies below are given without proof. Proofs can be constructed following similar arguments as in the case of the open queueing networks with blocking examined above. All services times are assumed to be exponentially distributed.

Closed queueing networks with arbitrary topologies: Consider a closed queueing network consisting of M nodes arbitrarily linked. Let m_i be the capacity of node i, i=1,2,...,M, N be the number of customers constantly circulating among the nodes of the network, and r_{ij} be the branching probability from node i to j. We have the following equivalencies

a) bbs-so and rs-fd are equivalent.
b) bbs-so and bbs-sno are equivalent for $N \le \min\{m_i+m_j,\ i,j=1,2,...,M \text{ such that } r_{ij}>0\}-1$.

Cyclic queueing networks: A cyclic queueing network consists of M nodes linked in series, as shown in figure 1.9. Let m_i be the capacity of node i, i=1,2,...,M, and N be the fixed number of customers constantly

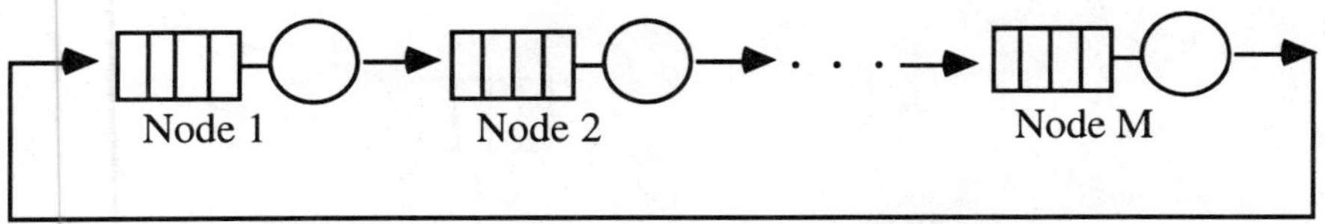

Figure 1.9: A cyclic network.

circulating through these nodes, each customer visiting nodes 1 to M in sequence. We have the following equivalencies

a) rs-fd and rs-rd are equivalent.
b) bbs-sno given node capacities m_i is equivalent to bas with node capacities m_i-1, i=1,2,...,M.

The following equivalencies hold when the cyclic queueing network consists of two nodes.

a) bbs-sno, bbs-so, rs-fd and rs-rd are equivalent.
b) bbs-sno (or bbs-so, or rs-fd, or rs-rd) with node capacities m_1 and m_2 is equivalent to bas with node capacities m_1-1 and m_2-1.

Central server model: The central server model is a closed queueing network consisting of nodes linked in a one-to-many (or many-to-one) configuration as shown in figure 1.10. Let m_i be the capacity of node i, i=0,1,...,M. We have the following equivalencies

a) bbs-so and bbs-sno are equivalent if $m_0=\infty$.
b) bbs-so and bbs-sno are equivalent if $m_i=\infty$, i=1,2,...,M.
c) rs-fd and rs-rd are equivalent if $m_i=\infty$, i=1,2,...,M.

1.4. The phase-type distribution

In this book, we will make extensive use of the phase-type distribution. This distribution is commonly used in performance studies of computer and communication systems. Its usefulness stems from the fact that it possesses the Markovian property, yet at the same time it can be used to represent (exactly or approximately) arbitrary distributions. It provides, therefore, the means of introducing arbitrary distributions in a Markovian queueing model without violating the Markovian property of the underlying process.

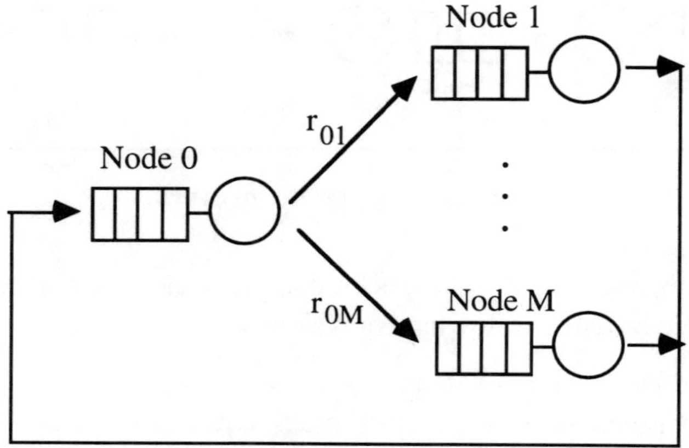

Figure 1.10: The central server model.

The Erlang distribution

Let us first consider the familiar Erlang distribution with k phases (or stages) shown in figure 1.11. The time spent in each phase is exponentially distributed with parameter μ. For simplicity, let us visualize this distribution as the service time in a node. A customer spends an exponentially distributed amount of time in the first phase, and then it proceeds to the second phase where it spends another exponentially distributed amount of time, and so on until it departs from the kth phase. The total time spent traversing k phases has a χ^2 distribution with 2k degrees of freedom. The representation of a service process or an inter-arrival process by an Erlang distribution may not necessarily mean that each phase has a physical interpretation. Let us now consider the Laplace transform, $f^*(s)$, of the density function of an Erlang distribution with k phases. We have

$$f^*(s) = f^*_{X_1}(s)\, f^*_{X_2}(s)\ \dots\ f^*_{X_k}(s),$$

where X_i is the time spent in the ith phase, i=1,2,...,k. Since this time is exponentially distributed we have

$$f^*(s) = \left(\frac{\mu}{s+\mu}\right)^k.$$

We can further generalize the above transformation by allowing the exponential distribution at each phase i to have a different mean μ_i, i=1,2,...,k. Then, we have

$$f^*(s) = \prod_{i=1}^{k} \left(\frac{\mu_i}{s+\mu_i}\right).$$

We observe that $f^*(s)$ has the form of a reciprocal polynomial of order k. Its roots $-\mu_i$, i=1,2,...,k, are negative real numbers. The number of roots (counting multiplicity) is equal to the number of phases. Also, each root can be seen as the rate out of a phase. Thus, any arbitrary distribution with a Laplace transform of the above type can be easily transformed to a sequence of exponential phases. This ingenious method of representing arbitrary distributions by a sequence of exponential phases was introduced by Erlang in his pioneering work on congestion in telephone systems.

The Coxian distribution

Let us now consider a family of arbitrary distributions whose Laplace transform $f^*(s)$ is rational. That is, it has the form

$$f^*(s) = \frac{\text{Polynomial in s of at most k order}}{\text{Polynomial in s of kth order}}.$$

Assuming that $-\mu_i$, i=1,2,...,k, are the roots of the denominator, we can expand $f^*(s)$ in partial functions. We have

$$f^*(s) = w_0 + \sum_{i=1}^{k} w_i \frac{\mu_i}{s+\mu_i},$$

where

$$\sum_{i=0}^{k} w_i = 1$$

and μ_i are all different. Also, it can be expanded as follows

Figure 1.11: The Erlang distribution.

$$f^*(s) = b_0 + \sum_{i=1}^{k} a_0 \ldots a_{i-1} b_i \prod_{j=1}^{i} \frac{\mu_j}{s+\mu_j},$$

where $a_i+b_i=1$, $i=1,2,...,k-1$, and $b_k=1$. The above expansion can be seen as the Laplace transform of the following distribution. There is a probability b_0 of zero service time and a probability a_0 of receiving an exponential service time with parameter μ_1. Then, there is a probability $b_1(=1-a_1)$ of departing, or a probability a_1 of receiving another exponential service time with parameter μ_2, and so on until the kth exponential service is received. This distribution is known as the *Coxian* distribution, and it is shown in figure 1.12 assuming $b_0=0$. This distribution is notated by the symbol C_k, where k is the number of phases. Notice that the Erlang distribution is a special case of the Coxian distribution.

It should be noted that in the above expansion, the roots μ_i, $i=1,2,...,k$, may not necessarily be real. In general, they may be complex numbers. We will not consider distributions with complex parameters. That is, we will assume that the roots μ_i, $i=1,2,...,k$, are always real.

We see, therefore, that any arbitrary distribution which has a rational Laplace transform may be expressed by a mixture of convolutions of exponential distributions arranged as shown above. This can be seen as a generalization of Erlang's idea.

The phase-type distribution

Cox also examined the case where the exponential phases are not visited in a sequential manner. That is, after phase i the next phase to enter is not only phase i+1 but any phase j, $j=1,2,...,k$, with probability a_{ij}, where $\sum_j a_{ij}=a_i$. He showed that such an arrangement has also a rational Laplace transform, and therefore it can be represented by the simpler configuration shown in figure 1.12. Such an arrangement of exponential phases is known as a *phase-type* distribution and it is notated as PH_k, where k is the

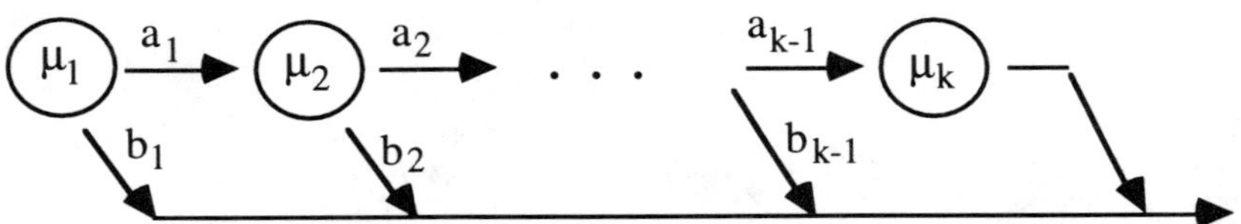

Figure 1.12: The Coxian distribution.

number of phases. The phase-type distribution has been used extensively due to its convenience in representing various types of distributions.

Let us consider an absorbing Markov process with states $\{1,2,...,k,k+1\}$. States 1 to k are associated with k exponential phases as in the Coxian distribution and state k+1 is an absorption state. Entering the absorbing state signifies the end of a time period distributed according to the phase-type distribution. The rate matrix **S** of this process involves all the transitions among the k states and out of these states into the absorbing state k+1. It can be divided, therefore, into the following different sub-matrices

$$\mathbf{S} = \begin{pmatrix} \mathbf{T} & \mathbf{T^o} \\ \mathbf{0} & 1 \end{pmatrix},$$

where **T** is the kxk matrix of the rates of transition among the k states, and $\mathbf{T^o}$ is a kx1 vector containing all the rates of transition out of these k states into the absorbing state k+1. In accordance with the usual way of writing down a rate matrix, each element gives the rate of transition from one state to another. The elements in each row i of matrix **S** are such that $T_{ij} \geq 0$, $i \neq j$. T_{ii} is set equal to the negative sum of all the elements in the row, that is $T_{ii} < 0$, $i=1,2,...,k$, and $\mathbf{Te}+\mathbf{T^o}=\mathbf{0}$, where **e** is a column vector of ones. In order to fully specify this Markov process we need to provide a probability vector of the initial distribution. That is, we need to know with what probability the process will start at phase i, i=1,2,...,k, k+1. Let this initial probability vector be $(\boldsymbol{\alpha},0)$ with $\boldsymbol{\alpha}$ being a 1xk row vector and $\boldsymbol{\alpha}\mathbf{e}=1$. Then, the distribution of the random variable representing the time until absorption in the above Markov Process is said to be of the phase-type with an $(\boldsymbol{\alpha},\mathbf{T})$ representation.

As an example of the phase-type distribution, let us consider the k-phase Coxian distribution discussed above. This distribution can be described in terms of a phase-type distribution as follows. The rate matrix **T**, which gives the rate transitions among the k phases, is

$$\mathbf{T} = \begin{pmatrix} -\mu_1 & a_1\mu_1 & & & \\ & -\mu_2 & a_2\mu_2 & & \\ & & -\mu_3 & a_3\mu_3 & \\ & & & \cdots & a_{k-1}\mu_{k-1} \\ & & & & -\mu_k \end{pmatrix},$$

and the other vectors are

$$\boldsymbol{\alpha} = (1,0,...,0) \text{ and } \mathbf{T^o} = [(1-a_1)\mu_1,(1-a_2)\mu_2,...,(1-a_k)\mu_k].$$

In general, the density function of a phase-type distribution with an $(\boldsymbol{\alpha},\mathbf{T})$ representation is

$$f(x) = \boldsymbol{\alpha} \exp(\mathbf{T}x)\mathbf{T^o}, \; x\geq 0,$$

and its Laplace transform is

$$f^*(s) = \boldsymbol{\alpha}(s\mathbf{I} - \mathbf{T})^{-1}\mathbf{T^o} \text{ for } res\geq 0.$$

The moments of f(x) are given by

$$E[X^n] = (-1)^n \, n! \; (\boldsymbol{\alpha} \, \mathbf{T}^{-n} \, \mathbf{e}), \; n\geq 1.$$

Some operations on phase-type distributions lead again to phase-type distributions. For instance, the convolution of two continuous phase-type distributions has a phase-type distribution. Also, a finite mixture of phase-type distributions has a phase-type distribution.

Some properties of the Coxian distribution

One of the most commonly used Coxian distributions is the one consisting of two phases, i.e. k=2. Let us consider a random variable X which has a two-phased Coxian distribution with parameters μ_1, μ_2 and a, as shown in figure 1.13. The probability density function of this distribution is

$$f(x) = (1-a)\mu_1 e^{-\mu_1 x} + a\left[\frac{\mu_1\mu_2}{\mu_2-\mu_1} e^{-\mu_1 x} + \frac{\mu_1\mu_2}{\mu_1-\mu_2} e^{-\mu_2 x}\right]$$

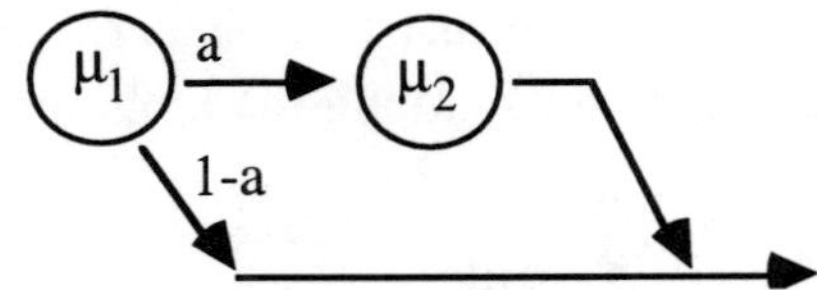

Figure 1.13: A Coxian distribution with two phases.

and its Laplace transform is

$$f^*(s) = (1\text{-}a)\frac{\mu_1}{s+\mu_1} + a\frac{\mu_1}{s+\mu_1}\frac{\mu_2}{s+\mu_2}$$

$$= \frac{\mu_1 s(1\text{-}a)+\mu_1\mu_2}{s^2+(\mu_1+\mu_2)s+\mu_1\mu_2}.$$

The first three moments can be obtained using the expression

$$\frac{d^n f^*(s)}{ds}\Big|_{s=0} = (-1)^n E(X^n).$$

We have

$$E(X) = \frac{1}{\mu_1} + \frac{a}{\mu_2}$$

$$E(X^2) = \frac{2(1\text{-}a)}{\mu_1^2} - \frac{2a\mu_1\mu_2\text{-}2a(\mu_1+\mu_2)^2}{\mu_1^2\mu_2^2}$$

$$E(X^3) = \frac{6(1\text{-}a)}{\mu_1^3} - \frac{12a\mu_1\mu_2(\mu_1+\mu_2)\text{-}6a(\mu_1+\mu_2)^3}{\mu_1^3\mu_2^3}.$$

From the above, we can calculate the variance and then the squared coefficient of variation c^2 of X. We have

$$c^2 = 1 - \frac{2a\mu_1[\mu_2\text{-}\mu_1(1\text{-}a)]}{(\mu_2+a\mu_1)^2}.$$

c^2 takes values in the range $[0.5,\infty)$. In general, it can be shown that for a C_k distribution, its c^2 takes values in the range $[1/k, \infty)$.

Under certain conditions, it can be shown that there is an equivalence between a Coxian distribution and a *hyperexponential* distribution. A hyperexponential distribution consists of k exponential phases, where $k \geq 2$, as shown in figure 1.14. Each phase i is associated with a probability p_i, where $\sum p_i=1$. The hyperexponential distribution is the distribution of the time spent in one of these k phases selected probabilistically. It is often notated as H_k, where k is number of phases. Its density function is

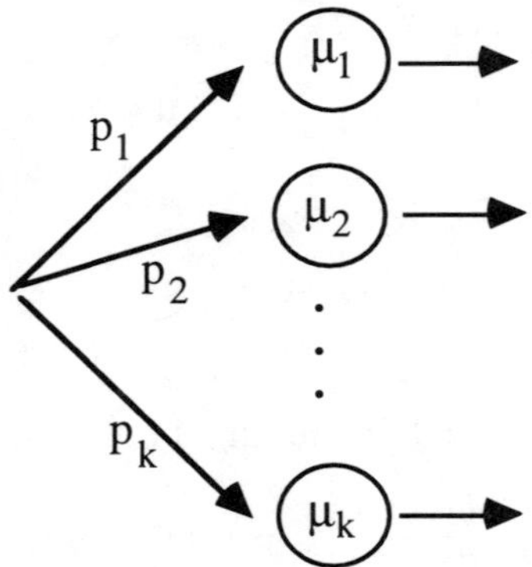

Figure 1.14: The hyperexponential distribution.

$$f(x) = \sum_{i=1}^{k} p_i \, \mu_i \, e^{-\mu_i x}, \quad x \geq 0,$$

and its Laplace transform is

$$f^*(s) = \sum_{i=1}^{k} p_i \frac{\mu_i}{s+\mu_i} \, .$$

Its c^2 takes values in the range $[1,\infty)$.

A Coxian distribution and a hyperexponential distribution are equivalent when the Laplace transform of their probability density function is the same. In general, under certain conditions, it can be shown that a Coxian distribution with k phases is equivalent to a hyperexponential distribution with k phases. The parameters of the exponential phases are the same. We demonstrate this equivalence for the case where k=2. Let us consider an H_2 with parameters $\lambda_1, \lambda_2, p_1, p_2$. Then, its Laplace transform $f^*_H(s)$ is

$$f^*_H(s) = \frac{(p_1\lambda_1+p_2\lambda_2)s+\lambda_1\lambda_2}{s^2+(\lambda_1+\lambda_2)s+\lambda_1\lambda_2} \, .$$

Now, let us consider a C_2 distribution with parameters μ_1, μ_2 and a. Its Laplace transform $f^*_C(s)$ is

$$f^*_C(s) = \frac{(1\text{-}a)\mu_1 s+\mu_1\mu_2}{s^2+(\mu_1+\mu_2)s+\mu_1\mu_2} \, .$$

We can obtain an equivalence by equating these two Laplace transforms. We observe that the two transforms are the same when $\mu_i=\lambda_i$, i=1,2, and $p_1\lambda_1+p_2\lambda_2=(1-a)\mu_1$. Now, given that we know the parameters of the H_2, the parameters of the equivalent C_2 are $\mu_i=\lambda_i$, i=1,2, and $a=p_2(\lambda_1-\lambda_2)/\lambda_1$. In order for a to be positive, we should have $\lambda_1>\lambda_2$. If this is not the case, then we can re-order the phases (and their corresponding branching probabilities) so that the above inequality holds. On the other hand, if the parameters of the C_2 distribution are given, the parameters of the equivalent H_2 are: $\lambda_i=\mu_i$, i=1,2, $p_2=1-p_1$, and $p_1=[\mu_1(1-a)-\mu_2]/(\mu_1-\mu_2)$. In order for p_1 to be positive we should have $\mu_1(1-a)>\mu_2$. This is guaranteed if the squared coefficient of variation of the Coxian distribution is greater than 1.

The generalized exponential distribution

We conclude this section with the generalized exponentia (GE) distribution. This is a special case of the H_2 distribution, and it is obtained by letting the rate of the second exponential phase go to infinity. (This is equivalent to considering a Coxian distribution with one phase, where $b_0>0$.) For convenience, the parameters of a GE distribution are defined as shown in figure 1.15. That is, with probability r there is an exponential service with rate ra and with probability 1-r there is zero service. We refer to this GE distribution as GE(r, ra). Let X be a random variable with a GE(r,ra) distribution. Then, its probability density function is

$$f(x) = (1-r)u(0) + r^2ae^{-rax},$$

where u(0)=1 for x=0. Its Laplace transform is

$$f^*(s) = 1-r + \frac{r^2a}{s+ra} \ ,$$

and its kth moment is

$$M_k = \frac{k!}{r^{k-1}a^k} \ .$$

The expected value of X is

$$r\frac{1}{ra} = \frac{1}{a}$$

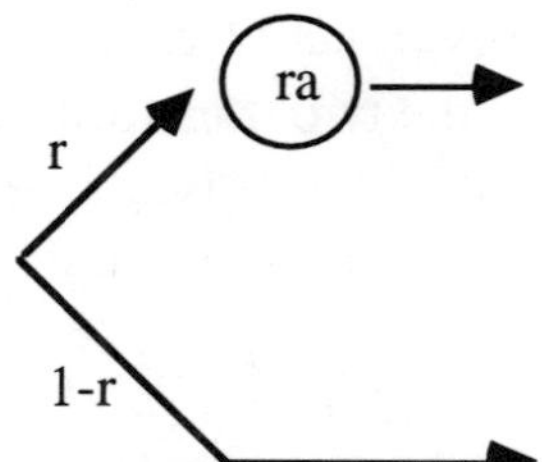

Figure 1.15: The generalized exponential distribution.

and its squared coefficient of variation is

$$c^2 = \frac{2}{r} - 1,$$

taking values in the region $[1,\infty)$. Thus, given a mean $1/\mu$ and a squared coefficient of variation c^2, the parameters of the GE distribution are $r=2/(1+c^2)$ and $a=\mu$.

Let us now consider an arrival process characterized by a GE(r,ra) distribution, and let us assume that an arrival has just occurred. Then, with probability r the next arrival will occur after an exponentially distributed time with rate ra, and with probability 1-r the next arrival will occur in zero time. That is, we may have multiple arrivals occurring at the same time. In view of this, the GE distribution can be seen as a bulk arrival distribution with geometrically distributed bulk and exponentially distributed bulk inter-arrival time. The probability that the bulk size is k is $(1-r)^{k-1}r$, and the mean bulk size is $1/r$.

The merging of N independent $GE(r_i,r_ia_i)$, $i=1,2,\ldots,N$, processes is a Poisson process with bulk arrivals and a mean inter-arrival time equal to $1/\sum r_ia_i$. The probability that a bulk comes from process i is $r_ia_i/\sum r_ia_i$, and the mean bulk size is

$$\sum \frac{r_ia_i}{\sum r_ia_i}\frac{1}{r_i} = \frac{\sum a_i}{\sum r_ia_i}.$$

In the special case where $r_i=r$, $i=1,2,\ldots,N$, the merged process is a $GE(r,r\sum a_i)$. Let c^2 and c_i^2 be the squared coefficient of variation of the merged process respectively of the ith process. We have, $c^2= (2/r)-1$, where $1/r$ is the mean bulk size. Thus,

$$c^2 = -1 + 2\frac{\Sigma a_i}{\Sigma r_i a_i}$$

$$= -1 + 2\frac{\Sigma a_i}{\Sigma \frac{2}{1+c_i^2} a_i}$$

$$= -1 + \left[\frac{\Sigma a_i/(1+c_i^2)}{\Sigma a_i}\right]^{-1}.$$

When splitting a GE(r,ra) process into N streams according to a Bernoulli filter $\{p_1,p_2,...,p_N\}$, the ith process is a $GE(r_i,r_ia_i)$, where $a_i=p_ia$, and $r_i=r/[r(1-p_i)+p_i]$. Moreover, if c^2 is the squared coefficient of variation of the GE(r,ra) process, then the squared coefficient of variation c_i^2 of the ith process is

$$c_i^2 = \frac{2}{r_i} - 1$$

$$= 1 + p_i(c^2-1).$$

Finally, let us consider the random sum of a GE(r,ra) distributed random variable over a geometric counter with parameter p. The Laplace transform $g^*(s)$ of the random sum is

$$g^*(s) = \frac{pf^*(s)}{1-(1-p)f^*(s)}.$$

Substituting for $f^*(s)$ we obtain

$$g^*(s) = \frac{p(1-r)}{p(1-r)+r} + \frac{r}{p(1-r)+r}\,\frac{pra/[p(1-r)+r]}{s+pra/[p(1-r)+r]}.$$

Thus, the random sum has a GE(r/[p(1-r)+r], pra/[p(1-r)+r]) distribution.

1.5. Approximating a general distribution by a Coxian distribution

As mentioned earlier, any distribution with a rational Laplace transform can be represented exactly by a Coxian distribution. The equivalent Coxian distribution, however, may have a large number of phases. From the practical point of view, therefore, it may be difficult to analyze a queueing model involving such a Coxian distribution. In this case, a Coxian distribution with a large number of phases is typically approximated by a Coxian distribution with a few phases. Coxian distributions with a large number of phases also arise in some of the approximation algorithms described in this book. In order to make these algorithms cpu efficient, these phase-type distributions are approximated by a Coxian distribution with only a small number of phases. Finally, it is often desirable to approximate a general service or arrival time distribution in a queueing model by a Coxian distribution in order to preserve the Markovian property. Below, we present some methods used to approximate an arbitrary distribution by a Coxian distribution for the case when the c^2 of the arbitrary distribution is greater than one and for the case when it is less than one.

$c^2 > 1$

Two-phase Coxian distributions are used frequently to approximate arbitrary distributions or phase-type distributions with a large number of phases whose squared coefficient of variation is greater than one. This is because a C_2 distribution is easily characterized by determining its three parameters μ_1, μ_2 and a, and it also gives rise to a smaller state space than the state space obtained if a Coxian distribution with more than two phases is used. There is no guarantee, however, that a two-phase Coxian distribution will produce a better fit than a Coxian distribution with more than two phases. There are several methods that can be used to approximate an arbitrary distribution by a C_2 distribution. Below, we discuss the method of moments, maximum likelihood estimation, and minimum distance estimation.

a. The method of moments

The probability density function of a C_2 and its first three moments were given in section 1.4. Now, let us assume that we know the first three exact or sample moments m_1, m_2, m_3 of the arbitrary distribution under study. Then, by equating these three moments with their counterpart moments of a C_2, we obtain a system of three equations which we can solve for μ_1, μ_2 and a. We have

$$m_1 = \frac{1}{\mu_1} + \frac{a}{\mu_2} \qquad (1.5.1)$$

$$m_2 = \frac{2(1\text{-a})}{\mu_1^2} - \frac{2a\mu_1\mu_2\text{-}2a(\mu_1+\mu_2)^2}{\mu_1^2\mu_2^2} \qquad (1.5.2)$$

$$m_3 = \frac{6(1\text{-a})}{\mu_1^3} - \frac{12a\mu_1\mu_2(\mu_1+\mu_2)\text{-}6a(\mu_1+\mu_2)^3}{\mu_1^3\mu_2^3} . \qquad (1.5.3)$$

From (1.5.1) we have

$$a = \frac{\mu_2}{\mu_1}(m_1\,\mu_1 - 1). \qquad (1.5.4)$$

Inserting (1.5.4) into (1.5.2) and (1.5.3) we obtain the equations

$$2m_1(\mu_1+\mu_2) - m_2\,\mu_1\mu_2 = 2,$$

$$6m_1(\mu_1+\mu_2)^2 - 6m_1\,\mu_1\mu_2 - 6(\mu_1+\mu_2) - m_3\,\mu_1^2\mu_2^2 = 0.$$

The above two equations can be simplified by setting $X=\mu_1+\mu_2$, and $Y= \mu_1\mu_2$. Solving for X and Y we have

$$Y = \frac{6m_1\text{-}3m_2/m_1}{(6m_2^2/4m_1) - m_3} \qquad (1.5.5)$$

$$X = \frac{1}{m_1} + \frac{m_2 Y}{2m_1} . \qquad (1.5.6)$$

Finally, we obtain μ_1 and μ_2 so that their sum is equal to X and their product Y. We have

$$\mu_1 = \frac{X + (X^2\text{-}4Y)^{1/2}}{2} \qquad (1.5.7)$$

$$\mu_2 = X - \mu_1. \qquad (1.5.8)$$

Given $m_1, m_2, m_3,$ the resulting C_2 is legitimate if the following condition is satisfied

$$3m_2^2 < 2m_1m_3. \qquad (1.5.9)$$

If this condition is not satisfied, one can choose the nearest acceptable value of the third moment or resort to a two moment approximation. In particular, let us assume that we know the mean m_1 and c^2 of an arbitrary distribution. Then, we can fit a C_2 distribution with parameters $\mu_1=2/m_1$, $\mu_2=1/m_1c^2$, and $a=1/2c^2$. These parameters can be used for an arbitrary distribution with $c^2>0.5$.

b. *Maximum likelihood estimation*

This method requires a sample of observations from the arbitrary distribution under study. Let this sample of size N be $x_1,x_2,\ldots,x_N$. Then the likelihood of the sample is

$$L(\theta) = \prod_{i=1}^{N} f(x_i),$$

where f(x) is the probability density of the fitted Coxian distribution, and $\theta=(\mu_1,\mu_2,a)$. The logarithm of the likelihood is

$$\log L(\theta) = \sum_{i=1}^{N} \log f(x_i).$$

The maximum likelihood estimate $\hat{\theta}=(\hat{\mu}_1,\hat{\mu}_2,\hat{a})$ is the value of θ which maximizes L or equivalently its logL subject to the constraints $\hat{\mu}_i\geq 0$, $i=1,2$, and $0\leq\hat{a}\leq 1$. The maximization can be carried out using a nonlinear optimization algorithm, since closed-form solutions of the likelihood equations do not exist.

c. *Minimum distance estimation*

This method is based on the idea of estimating the unknown parameters of the fitted distribution by minimizing the distance between the fitted distribution and the observed one. It requires a sample of observations from the arbitrary distribution under study.

Let F(x) be the distribution function of the arbitrary general distribution. Let $F_\theta(x)$ be the distribution function that will be used to approximate F(x). Consider a random sample $x_1,x_2,\ldots,x_N$ from the population with distribution function F(x). Let $F_N(x)$ denote the empirical distribution function, i.e.

$$F_N(x) = \frac{1}{N}\sum_{i=1}^{N} I(x_i\leq x).$$

Parameter θ is estimated by selecting those values such that the distance between F_N and F_θ is minimized. The computing formula for the Cramer-Von Miser distance between F_N and F_θ is

$$d^2 = \frac{1}{12N} + \sum_{i=1}^{N} [F_\theta(x_{(i)}) - \frac{i-.5}{N}],$$

where $x_{(i)}$ is the ith order statistic and $F_\theta(x)$ is a C_2. The above minimization problem does not have a closed from solution. It can be solved using a non-linear optimization algorithm.

Comparisons among the above three methods have shown that when fitting a C_2 distribution to an arbitrary distribution, the method of moments gives better results than the maximum likelihood or minimum distance methods, if the exact moments of the original distribution are known. However, in the case where the original distribution is not known, it has been observed that the maximum likelihood distribution or the minimum distance gives better results. Further, the method of minimum distance is robust in the presence of outliers.

We note that the method of moments is an easy method to use as opposed to the other two methods which require nonlinear optimization. Hence, it is a convenient method to incorporate in an approximation algorithm.

$c^2 < 1$

Assuming that only the mean m_1 and the coefficient of variation c^2 of the arbitrary distribution are known, we can fit the following generalized Erlang distribution. (A generalized Erlang distribution is an Erlang distribution in which after the first exponential phase, the service may end with probability 1-a or it may continue through the remaining k-1 phases with probability a.) The number of stages k should be such that

$$\frac{1}{k} \leq c^2 \leq \frac{1}{k-1} .$$

Once k is fixed, the other parameters μ and a can be obtained from the expressions

$$1-a = \frac{2kc^2+k-2-(k^2+4-4kc^2)^{1/2}}{2(c^2+1)(k-1)}$$

$$\mu = \frac{1+(k-1)a}{m_1} .$$

For the range of $0.5 \leq c^2 \leq 1$, it is also possible to use the two moment approximation given above, i.e. $\mu_1 = 2/m_1$, $\mu_2 = 1/m_1c^2$, and $a = 1/2c^2$.

1.6. References

The material given in section 1.3 for open queueing networks with blocking is due to Onvural and Perros (1986). The equivalencies for closed queueing networks with blocking given in section 1.3 are due to Balsamo and De Nitto Personè (1991), and Onvural (1987). The Coxian distribution described in section 1.4 is due to Cox (1955). In principle, we may say that a Coxian distribution may be used to approximate any arbitrary distribution. (See Newman and Reddy (1979) for a rational approximation to e^X and related functions.) The phase-type distribution is due to Neuts (1981). The generalized exponential (GE) distribution is due to Kouvatsos (1988). See also Xenios (1989). In section 1.5, the expressions for the three moment approximation are due to Altiok (1985), and the expressions for the two moment approximation are due to Marie (1978). Finally, the expressions for fitting a generalized Erlang distribution when $c^2<1$ are due to Sauer and Chandy (1975). Further references related to the topics presented in this Chapter are given below.

Blocking mechanisms

The mechanisms blocking-after-service and blocking-before-service have been used in models of production systems (see Altiok (1982) and Gershwin (1987)). The blocking-after-service mechanism has been used in a model of a disk I/O system (see Perros (1981)). The repetitive-service mechanism with fixed destination has been used in models of telecommunication systems (see Caseau and Pujolle (1979)), and the repetitive-service mechanism with random destination has been used in a model of a flexible manufacturing system (see Yao and Buzacott (1987)). Other blocking mechanisms have been considered by Altiok and Ranjan (1988), Hordijk and van Dijk (1981), Latouche and Neuts (1980), Lavenberg (1978), Le Ny (1980) and (1983), van Dijk (1991a) and (1991b), and Yao and Buzacott (1985).

Equivalencies between blocking mechanisms

Comparisons between the three blocking mechanisms discussed in section 1.3 for open exponential queueing networks with blocking have also been reported in Altiok and Stidham (1982), Bocharov and Albores (1980), and Caseau and Pujolle (1979). Sakasegawa and Kawashima (1992) reported on equivalencies in non-Markovian tandem configurations. Finally, a survey of product-form closed queueing networks with blocking and their equivalencies is given in Balsamo and De Nitto Personè (1991).

The phase-type distribution

Phase-type distributions have been used extensively in a variety of forms. For a detailed discussion on the phase-type distribution see Neuts(1981). Whitt (1984) used the hyperexponential distribution to approximate the behaviour of arrival processes in queueing systems. Botta and Harris (1986) introduced the generalized hyperexponential distribution as a linear combination of exponential distributions. Finally, Johnson and Taaffe (1986) used a particular phase-type structure for approximating arbitrary distributions.

Bibliography

Altiok, T., Approximate analysis of exponential tandem queues with blocking, *Eur. J. Oper. Res.* **11** (1982) 390-398.

Altiok, T., On the phase-type approximations of general distributions, *IIE Trans.* **17** (1985) 110-116.

Altiok, T. and Ranjan, R., *Multi-stage, pull-type production/inventory systems with set-ups and (R,r) control policies*, Tech. Rept. 88-108, Industrial Engineering Dept., Rutgers Univ., 1988.

Altiok, T. and Stidham, S., A note on transfer lines with unreliable machines, random processing times and finite buffers, *IIE Trans.* (1982) 125-127.

Balsamo, S. and De Nitto Personè, V., Closed queueing networks with finite capacities: blocking types, product-form solution and performance indices, *Perf. Eval.* **12** (1991) 85-102.

Balsamo, S., and De Nitto Personè, V., *A survey of product-form queueing networks with blocking and their equivalences*, Tech. Rept., Dipartmento di Ingegneria Elettronica, Universita di Roma, 1992. To appear in *Ann. Oper. Res.*

Bocharov, P. P. and Albores, F. K., On two-stage exponential queueing system with internal losses or blocking, *Prob. Control Infor. Theory* **9** (1980) 365-379.

Botta, R. F. and Harris, C. M., Approximation with generalized hyperexponential distributions: weak convergence results, *Queueing Systems*, **2** (1986) 169-190.

Caseau, P. and Pujolle, G., Throughput capacity of a sequence of queues with blocking due to finite waiting room, *IEEE Trans. Soft. Eng.* **5** (1979) 631-642.

Cox, D. R., A use of complex probabilities in the theory of stochastic processes, *Proc. Cambridge Philos. Soc* **15** (1955) 313-319.

Gershwin, S. B., An efficient decomposition method for the approximate evaluation of tandem queues with finite storage space and blocking, *Oper. Res.* **35** (1987) 291-305.

Hordijk, A. and van Dijk, N., Networks of queues with blocking, in: Kylstra, Ed., *Performance '81*, (North-Holland, Amsterdam, 1981) 51-65.

Johnon, M. A. and Taaffe, M. R., Matching moments to phase distributions: mixtures of Erlang distributions of common order, *Commun. Statist.-Stochastic Models*, **5** (1989) 711-743.

Kouvatsos, D. D., A maximum entropy analysis of the G/G/1 queue at equilibrium, *J. Op. Res. Soc.* **39** (1988)

Latouche, G. and Neuts, M. F., Efficient algorithmic solutions to exponential tandem queues with blocking, *SIAM J. of Alg. Disc. Meth.* **1** (1980) 93-106.

Lavenberg, S. S., Stability and maximum departure rate of certain open queueing networks having finite capacity constraints, *RAIRO Informatique/Computer Science* **12** (1978) 353-370.

Marie, R., *Modelisation par reseaux de files d'attende*, Ph.D. Thesis, L'Université de Rennes, France, 1978.

Le Ny, L.-M., Étude analytique de reseaux de files d'attende multiclasses à routage variable, *RAIRO Recherche Operationnelle/Oper. Res.* **14** (1980) 331-347.

Le Ny, L.-M., Forme produit pour des reseaux multiclasses à routage dynamique, *Annales Scientifiques de l'Universitè de Clermont-Ferrand II* **76** (1983) 17-34.

Neuts, M. F., *Matrix-geometric solutions in stochastic models - an algorithmic approach*, (John Hopkins University Press, 1981).

Newman, D. J. and Reddy, A. R., Rational approximation to e^x and to related functions, *J. of Approximation Theory* **25** (1979) 21-30.

Onvural, R. O., *Closed queueing network with finite buffers*, Ph.D. thesis, North Carolina State University, 1987.

Onvural, R. O. and Perros, H. G., On equivalencies of blocking mechanisms in queueing networks with blocking, *Oper. Res. Lett.* **5** (1986) 293-297.

Perros, H. G., A symmetrical exponential open queue network with blocking and feedback, *IEEE Trans. Soft. Eng.* **7** (1981) 395-402.

Sakasegawa, H. and Kawashima, T., *On the equivalence of three types of blocking in non-Markovian tandem queues*, Tech. Rept., Institute of Socio-Economic Planning, Univ. of Tsukuba, 1992.

Sauer, C. H. and Chandy, K. M., *Approximate analysis of central server models*, Tech. Rept., IBM T.J. Watson Research Center, Yorktown Heights, 1975.

van Dijk, N. M., On 'stop=repeat' servicing for non-exponential queueing networks with blocking, *J. Appl. Probab.* **28** (1991a) 159-173.

van Dijk, N. M., 'Stop=recirculate' for exponential product-form queueing networks with departure blocking, *Oper. Res. Lett.* **10** (1991b) 343-351.

Xenios, N. P., *General queueing networks with blocking*, Ph.D. thesis, University of Bradford, 1989.

Yao, D. D. and Buzacott, J. A., Modelling a class of state-dependent routing in flexible manufacturing systems, *Ann. Oper. Res.* **3** (1985) 153-167.

Yao, D. D. and Buzacott, J. A., Modeling a class of flexible manufacturing systems with reversible routing, *Oper. Res.* **35** (1987) 87-93.

Whitt, W., On approximations for queues, III: mixtures of exponential distributions, *AT&T Bell Labs Tech. J.*, **64** (1984) 163-175.

CHAPTER 2

Numerical Methods For Queueing Networks With Blocking

2.1. Introduction

Numerical methods provide an alternative approach to the analysis of queueing systems. They are particularly useful in cases where it is not possible to obtain an analytic solution for the queueing system under study. The numerical methods we refer to are basically equation solving techniques. The queueing system under study is first formulated as a continuous time Markov process with discrete states, and subsequently its steady-state probability vector is calculated using an equation solving technique. Many useful performance measures regarding the system under study can be extracted from the steady-state probability vector. The applicability of these numerical methods is limited by memory and cpu constraints. Computer memory requirements become severe when the state space of the Markov process under study becomes very large. Also, setting-up the rate matrix and calculating the steady-state probability vector may be cpu intensive.

In general, these numerical methods are divided into two categories, namely, *direct* and *iterative* methods. In principle, direct methods yield exact results in a finite number of arithmetic operations. In practice, they produce approximate results due to round-off errors. In fact, errors resulting from rounding-off and also from instability may lead to poor results. Examples of direct methods are the Gaussian elimination method and its variations. Iterative methods start with an initial approximation and then proceed with an appropriate algorithm to obtain successively better approximations. The results are only approximate and the issues of convergence and accuracy are crucial. The advantages of iterative methods are simplicity and uniformity of the arithmetic operations. They are also relatively insensitivity to round-off errors, and they lend themselves to compact storage schemes. There also exist many methods which exploit the special structure of the rate matrix of a Markov process. For instance, the matrix-geometric method can be employed to analyze Markov processes which have a certain block structure, and various decomposition methods can be used for *nearly completely decomposable* Markov processes.

In this Chapter, we will describe some of the commonly used numerical methods to solve for the steady-state probability vector of a Markov process. In the following section, we review some basic concepts related to continuous time Markov processes with discrete states, and we discuss how they can be analyzed using equation solving techniques. In section 2.3, an example of rate matrix generation is given. In section 2.4, we briefly discuss various direct and iterative numerical methods for solving systems of linear equations, and in section 2.5 we describe the matrix-geometric procedure. Finally, relevant references are given in section 2.6. We note that matrices will be denoted by capital letters in bold, and vectors will be denoted by small letters in bold. In order to simplify the notation we will assume that all probability vectors are row vectors.

2.2. Continuous time Markov processes with discrete states

In this section, we consider a continuous time Markov process with discrete states. Such a process will be referred to as a *Markov process*. A discrete time Markov process with discrete states will be referred to as a *Markov chain*. A Markov process spends an exponentially distributed period of time at each state, which is independent of the past behaviour of the process. Let $X(t)$ denote the state occupied at time t. A Markov process is in principle defined when we have a set of functions $p_{ij}(s,t)$ with the interpretation

$$p_{ij}(s,t) = \text{Prob}(X(t)=j \mid X(s)=i).$$

This is the probability that the process is in state j at time t given that the process was in state i at time s. In the case where the transitions depend only on the difference $\tau=t-s$, the Markov process is said to be *homogeneous*. In this case, we have

$$p_{ij}(\tau) = \text{Prob}(X(s+\tau)=j \mid X(s)=i),$$

where

$$\sum_j p_{ij}(\tau) = 1, \text{ for all values of } \tau.$$

Now, let us consider the limiting case of $p_{ij}(t,t+\Delta t)$ as $\Delta t \to 0$ for $i \neq j$, where Δt is a small finite number. $p_{ij}(t,t+\Delta t)$ is the probability that the process will move from state i to state j in the interval $[t,t+\Delta t)$. As Δt becomes very small, the probability of a transition during this time $[t,t+\Delta t)$ becomes very small as well, i.e.

$$p_{ij}(t, t+\Delta t) \to 0 \text{ as } \Delta t \to 0.$$

In the case where i=j, as Δt becomes very small, the probability of staying in the same state i increases. That is,

$$p_{ii}(t, t+\Delta t) \to 1 \text{ as } \Delta t \to 0.$$

Based on the above observations, we introduce the limits

$$\lim_{\Delta t \to 0} \frac{p_{ij}(t,t+\Delta t) - 0}{\Delta t} = q_{ij}(t), \ i \neq j$$

$$\lim_{\Delta t \to 0} \frac{p_{ii}(t,t+\Delta t) - 1}{\Delta t} = q_{ii}(t).$$

The quantities $q_{ij}(t)$ can be interpreted as the rate of transition from state i to state j at time t. The rate of transition is defined as the mean number of transitions per unit time. The above equations permit us to express a probability $p_{ij}(t,t+\Delta t)$ in terms of its transition rate $q_{ij}(t)$. That is,

$$p_{ij}(t,t+\Delta t) = q_{ij}(t)\Delta t + o(\Delta t), \ i \neq j. \quad (2.2.1)$$

This can be interpreted as follows. The probability that a transition occurs from state i to state j in Δt is equal to the rate of transitions during Δt plus the probability of observing multiple transitions which is of the order $o(\Delta t)$. That is, the probability of observing multiple transitions tends to zero faster than Δt. When i=j we have

$$p_{ii}(t,t+\Delta t) = 1 + q_{ii}(t)\Delta t + o(\Delta t). \quad (2.2.2)$$

From the above equation we have

$$
\begin{aligned}
-q_{ii}(t)\Delta t + o(\Delta t) &= 1- p_{ii}(t,t+\Delta t) \\
&= \sum_{j\neq i} p_{ij}(t,t+\Delta t) \\
&= \sum_{j\neq i} q_{ij}(t)\Delta t + o(\Delta t)
\end{aligned}
$$

or

$$q_{ii}(t) = - \sum_{j\neq i} q_{ij}(t).$$

Now, let us define the matrix $\mathbf{Q}(t)$, whose (i,j)th element is $q_{ij}(t)$. This matrix is called the *infinitesimal generator matrix* or the *transition rate matrix*, or simply the *rate matrix* for a Markov process. Expressions (2.2.1) and (2.2.2) can be written in matrix form as follows

$$\mathbf{Q}(t) = \lim_{\Delta t \to 0} \frac{\mathbf{P}(t) - \mathbf{I}}{\Delta t} , \qquad (2.2.3)$$

where $\mathbf{P}(t)=(p_{ij}(t,t+\Delta t))$ is the *transition probability matrix* and $\mathbf{I}$ is the identity matrix, i.e. a matrix whose diagonal elements are 1 and the non-diagonal elements are zero. From (2.2.3), we can write

$$\mathbf{P}(t) = \mathbf{Q}(t)\Delta t + \mathbf{I} , \qquad (2.2.4)$$

where Δt is small enough such that no more than one transition can take place during Δt. However, expression (2.2.4) allows us to discretize the Markov process $\mathbf{P}(t)$ and express it as a Markov chain with a unit time interval Δt.

Now, let $x_i(t)$ be the probability that the system is in state i at time t, i.e. $x_i(t)=\text{Prob}(X(t)=i)$. Then, we have

$$x_i(t+\Delta t) = x_i(t) \Big(1-\sum_{j\neq i} p_{ij}(t,t+\Delta t)\Big) + \sum_{j\neq i} x_j(t) p_{ji}(t,t+\Delta t)$$

or

$$x_i(t+\Delta t) = x_i(t) \Big(1-\sum_{j\neq i} q_{ij}(t)\Delta t\Big) + \sum_{j\neq i} q_{ji}(t) x_j(t)\Delta t + o(\Delta t)$$

or

$$\frac{x_i(t+\Delta t)-x_i(t)}{\Delta t} = -x_i(t)\sum_{j\neq i} q_{ij}(t) + \sum_{j\neq i} q_{ji}(t)x_j(t) + o(\Delta t)$$

$$= x_i(t)q_{ii}(t) + \sum_{j\neq i} q_{ji}(t)x_j(t) + o(\Delta t)$$

$$= \sum_{j} q_{ji}(t)x_j(t) + o(\Delta t).$$

Thus,

$$\lim_{\Delta t\to 0} \frac{x_i(t+\Delta t)-x_i(t)}{\Delta t} = \dot{x}_i(t) = \sum_j q_{ji}(t)x_j(t)$$

or, in matrix form

$$\dot{\mathbf{x}}(t) = \mathbf{x}(t)\,\mathbf{Q}(t), \qquad (2.2.5)$$

where $\mathbf{x}(t)$ is a probability vector whose ith position is $x_i(t)$.

For homogeneous Markov processes in steady-state, we simply have

$$\mathbf{xQ} = \mathbf{0} \qquad (2.2.6)$$

where $\mathbf{x}$ is the stationary probability vector whose ith element x_i is the steady-state probability that the system is in state i. $\mathbf{x}$ can be obtained from (2.2.6) and the normalizing condition $\sum x_i=1$, using equation solving techniques. Alternatively, $\mathbf{x}$ may be obtained as follows. We can multiply (2.2.6) by Δt and then add $\mathbf{x}$ on both sides of the equation, so that we have $\mathbf{xQ}\Delta t+\mathbf{x}=\mathbf{x}$, or

$$\mathbf{xP} = \mathbf{x} \qquad (2.2.7)$$

where $\mathbf{P}=\mathbf{Q}\Delta t+\mathbf{I}$. The quantity Δt has to be chosen so that $\Delta t \leq 1/\max_i|q_{ii}|$ in order for $\mathbf{P}$ to be a stochastic matrix. The value of $\Delta t=0.99/\max_i|q_{ii}|$ appears to be suitable in most cases. In this book we will only deal with homogeneous Markov processes in steady-state.

A queueing network with blocking, under appropriate assumptions, can be formulated as a Markov process. The stationary probability vector can be obtained from (2.2.6) or (2.2.7) using numerical methods for linear systems of equations. In general, this approach involves the following three steps: a) generation of the state space, b) generation of the rate matrix **Q**, and c) numerical solution of (2.2.6) or (2.2.7) in order to calculate the stationary probability vector. Steps a) and b) are presented below through an example. Numerical methods for solving linear systems are discussed in the subsequent sections.

2.3. An example of rate matrix generation

Let us consider the two-node queueing network with blocking shown in figure 2.1. Each node consists of a queue served by a single server. The first node has an infinite capacity and the second node has a finite capacity of size m. The service time at each node is exponentially distributed. External arrivals at the first node follow a Poisson distribution with the rate λ. Let μ_1, μ_2 be the service rate at the first and at the second server respectively. Customers in each node are served in a FIFO manner.

We assume blocking-after-service. That is, the server at the first node becomes blocked when a customer completes its service at a time the second node is full. The server remains blocked, and it cannot serve any other customer in its queue, until a departure occurs from node 2. This is probably the simplest queueing network with blocking, yet as it will be seen in the following Chapter, its analysis is not easy.

The state of this queueing network can be described by the vector (n_1,n_2), where n_1 indicates the number of units in the first node and n_2 indicates the number of units in the second node and the blocking status. We have, $n_1=0,1,...$, and $n_2=0,1,...,m,m+1$. $n_2=m+1$ indicates that the second node is full and that the server at the first node is blocked. Thus, any state $(i,m+1)$, $i=1,2,...$, indicates that the first server is blocked and that there are i units in the first node. Of these i units, one has received its service but it is blocked from entering node 2, and the

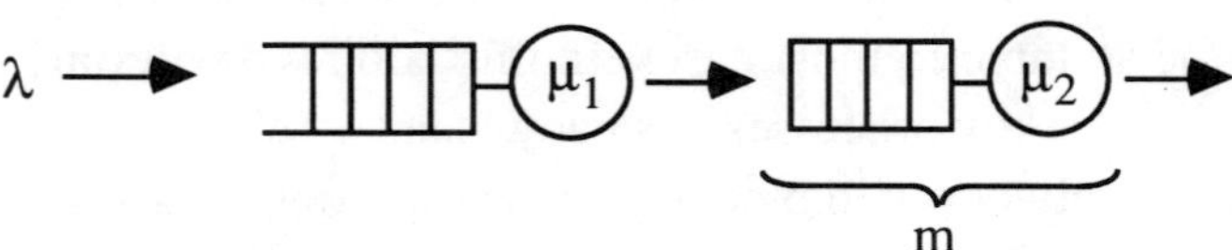

Figure 2.1: A two-node open queueing network with blocking.

remaining i-1 are waiting to be served. Using this state description, we can order the states as follows:

$$\{(0,0),(0,1),\ldots,(0,m),(1,0),(1,1),\ldots,(1,m),$$
$$(1,m+1),\ldots,(i,0),(i,1),\ldots,(i,m),(i,m+1),\ldots\}.$$

We note that above states were ordered *lexicographically*. Described in simple terms, the states (n_1,n_2) were ordered by first fixing the value of n_1 and subsequently considering all the feasible values of n_2 in an ascending order. The value of n_1 was then increased by 1 and the above procedure was repeated.

The generation of the rate matrix **Q** involves the creation of a matrix containing all the possible transitions from each state. For instance, the rate matrix **Q** of the above two-node network, assuming m=2, is

	(0,0)	(0,1)	(0,2)	(1,0)	(1,1)	(1,2)	(1,3)	(2,0)	(2,1)	(2,2)	(2,3)	(3,0)	(3,1)	(3,2)	(3,3)
(0,0)	*			λ											
(0,1)	μ_2	*			λ										
(0,2)		μ_2	*			λ									
(1,0)		μ_1		*				λ							
(1,1)			μ_1	μ_2	*				λ						
(1,2)					μ_2	*	μ_1			λ					
(1,3)			μ_2				*				λ				
(2,0)					μ_1			*				λ			
(2,1)						μ_1		μ_2	*				λ		
(2,2)									μ_2	*	μ_1			λ	
(2,3)						μ_2					*				λ
.															
.															
.															

The horizontal and vertical lines have been drawn to further subdivide the matrix into sub-matrices, or *blocks*. We note that the matrix has a block tri-diagonal structure. That is, the blocks are arranged along a diagonal band consisting of the diagonal, the line immediately above the diagonal and the line immediately below the diagonal. For each row, the diagonal element represented by a * is the negative sum of all the other elements in the same row. For instance, the diagonal element of the first row is $-\lambda$, the diagonal element of the second row is $-(\mu_2+\lambda)$, etc. In view of this, the sum of all the elements in each row is zero.

The rate matrix of a queueing network is usually a sparse matrix with very small intensity (ratio of the number of non-zero elements over the total number of elements). Therefore, a significant amount of memory can be saved if we only store the non-zero elements. There are various compact storage schemes that can be used. The simplest scheme is to set up a linked list, where each element of the list contains the fields: column number, row number, value of the non-zero element, and a pointer to the next element in the list. An improvement over this approach would be to keep a separate linked list for each column of the rate matrix. The elements of each of these lists will only contain the following three fields: row number, value of the non-zero element, and a pointer to the next element in the list. Each linked list will have an appropriate header containing the fields: column number, a pointer to the first element of the list, and a pointer to the header of the next column. More savings are possible if the transition rates in the rate matrix repeat themselves. These transition rates are stored in a table. Then, instead of storing each transition rate separately, we simply store a pointer to the appropriate entry in the table of transition rates. Clearly, the more elaborate the storage scheme is, the more involved it is to access the elements of the rate matrix.

2.4. Direct and iterative numerical methods

In this section, we review some of the direct and iterative numerical methods that can be used to solve the system of linear equations $\mathbf{xQ=0}$ or $\mathbf{Q}^T\mathbf{x}^T\mathbf{=0}$. In order to simplify the notation, we set $\mathbf{x}^T\mathbf{=y}$. Thus, $\mathbf{Q}^T\mathbf{x}^T\mathbf{=0}$ will be written as $\mathbf{Q}^T\mathbf{y=0}$. Let q_{ij} be the (i,j)th element of matrix $\mathbf{Q}$, and let n be the total number of states. (If $\mathbf{Q}$ has an infinite number of states, then it is truncated to a finite number of states.)

One of the well-known direct methods for solving $\mathbf{Q}^T\mathbf{y=0}$ is Gaussian elimination. Let us first consider the LU decomposition of the matrix $\mathbf{Q}$. We have $\mathbf{Q=LU}$, where

$$\mathbf{L} = \begin{pmatrix} l_{11} & & & \\ l_{21} & l_{22} & & \\ & \cdots & & \\ l_{n1} & l_{n2} & \cdots & l_{nn} \end{pmatrix}, \quad \mathbf{U} = \begin{pmatrix} u_{11} & u_{12} & \cdots & u_{1n} \\ & u_{22} & \cdots & u_{2n} \\ & & \cdots & \\ & & & u_{nn} \end{pmatrix}.$$

The unknowns l_{ij} and u_{ij} can be obtained from the system of linear equations **Q=LU**. We have

$$\sum_{k=1}^{j} l_{ik}u_{kj} = q_{ij}, \text{ if } i > j \tag{2.4.1}$$

$$\sum_{k=1}^{i} l_{ik}u_{kj} = q_{ij}, \text{ if } i \leq j \tag{2.4.2}$$

In order to reduce the number of unknowns, we set $l_{ii}=1$, $i=1,2,...,n$. Expression (2.4.2) can then be re-written as

$$\sum_{k=1}^{i-1} l_{ik}u_{kj} + u_{ij} = q_{ij}$$

or

$$u_{ij} = q_{ij} - \sum_{k=1}^{i-1} l_{ik}u_{kj} . \tag{2.4.3}$$

Similarly, expression (2.4.1) can be re-written as

$$\sum_{k=1}^{j-1} l_{ik}u_{kj} + l_{ij}u_{jj} = q_{ij}$$

or

$$l_{ij} = \frac{1}{u_{jj}} [q_{ij} - \sum_{k=1}^{j-1} l_{ik}u_{kj}]. \tag{2.4.4}$$

As an example, let us assume that

$$\mathbf{Q} = \begin{pmatrix} -4 & 2 & 2 \\ 2 & -3 & 1 \\ 1 & 1 & -2 \end{pmatrix}.$$

Then, matrices **L** and **U** are

$$\mathbf{L} = \begin{pmatrix} 1 & & \\ -1/2 & 1 & \\ -1/4 & -3/4 & 1 \end{pmatrix}, \qquad \mathbf{U} = \begin{pmatrix} -4 & 2 & 2 \\ & -2 & 2 \\ & & 0 \end{pmatrix}.$$

Let us now return to our system of linear equations $\mathbf{Q}^T\mathbf{y}=\mathbf{0}$. Using the LU decomposition of **Q** we have $(\mathbf{LU})^T\mathbf{y}=\mathbf{0}$, or $\mathbf{U}^T\mathbf{L}^T\mathbf{y}=\mathbf{0}$. Now, set $\mathbf{L}^T\mathbf{y}=\mathbf{z}$ and then solve $\mathbf{U}^T\mathbf{z}=\mathbf{0}$. Once **z** has been calculated, we can solve the system $\mathbf{L}^T\mathbf{y}=\mathbf{z}$ in order to obtain **y**. However, the system $\mathbf{U}^T\mathbf{z}=\mathbf{0}$ cannot be solved. In view of this, we replace one of the equations in the original system $\mathbf{xQ}=\mathbf{0}$, by the normalizing equation $\mathbf{xe}=1$. The resulting system is $\overline{\mathbf{Q}}^T\mathbf{y}=\mathbf{e}_n$, where $\overline{\mathbf{Q}}$ is identical to **Q** except that the last column is replaced by 1's, and $\mathbf{e}_n$ is a column vector whose nth row is 1 and all other rows are 0. Let $\overline{\mathbf{Q}}=\overline{\mathbf{L}}\,\overline{\mathbf{U}}$. Then, we have $(\overline{\mathbf{L}}\,\overline{\mathbf{U}})^T\mathbf{y}=\mathbf{e}_n$, or $\overline{\mathbf{U}}^T\overline{\mathbf{L}}^T\mathbf{y}=\mathbf{e}_n$. As before, we set $\overline{\mathbf{L}}^T\mathbf{y}=\mathbf{z}$, and then solve $\overline{\mathbf{U}}^T\mathbf{z}=\mathbf{e}_n$. Since $\mathbf{e}_n$ is all zeros, except the nth row, the solution for **z** is $\mathbf{z}=(0,\ldots,0,1/u_{nn})$. Thus, $\overline{\mathbf{L}}^T\mathbf{y}=\mathbf{z}$ becomes

$$\begin{pmatrix} 1 & l_{21} & \ldots & l_{n1} \\ & 1 & \ldots & l_{n2} \\ & & \ldots & \\ & & & 1 \end{pmatrix} \begin{pmatrix} y_1 \\ y_2 \\ . \\ y_n \end{pmatrix} = \begin{pmatrix} 0 \\ 0 \\ . \\ 1/u_{nn} \end{pmatrix}$$

which can be solved by backward substitution. Following the same example as before, we have

$$\overline{\mathbf{Q}} = \begin{pmatrix} -4 & 2 & 1 \\ 2 & -3 & 1 \\ 1 & 1 & 1 \end{pmatrix}$$

and

$$\bar{\mathbf{L}} = \begin{pmatrix} 1 & & \\ -1/2 & 1 & \\ -1/4 & -3/4 & 1 \end{pmatrix}, \quad \bar{\mathbf{U}} = \begin{pmatrix} -4 & 2 & 1 \\ & -2 & 3/2 \\ & & 19/8 \end{pmatrix}.$$

Thus, **y** can be obtained from the system

$$\begin{pmatrix} 1 & -1/2 & -1/4 \\ & 1 & -3/4 \\ & & 1 \end{pmatrix}\begin{pmatrix} y_1 \\ y_2 \\ y_3 \end{pmatrix} = \begin{pmatrix} 0 \\ 0 \\ 8/19 \end{pmatrix}.$$

Iterative methods usually derive a recursion formula from which the approximate values of the unknowns are obtained iteratively. Let us consider the system $\mathbf{Q}^T\mathbf{y}=\mathbf{0}$. Matrix $\mathbf{Q}^T$ is decomposed into a lower sub-matrix **B**, a diagonal sub-matrix **D**, and an upper sub-matrix **H** so that $\mathbf{Q}^T= \mathbf{B}+\mathbf{D}+\mathbf{H}$, where

$$\mathbf{B} = \begin{pmatrix} 0 & 0 & 0 & \dots & 0 & 0 \\ q_{12} & 0 & 0 & \dots & 0 & 0 \\ q_{13} & q_{23} & 0 & \dots & 0 & 0 \\ & \cdot & \cdot & \cdot & & \\ & & \cdot & \cdot & \cdot & \\ q_{1n} & q_{2n} & q_{3n} & \dots & q_{n-1,n} & 0 \end{pmatrix},$$

$$\mathbf{D} = \begin{pmatrix} q_{11} & 0 & 0 & \dots & 0 & 0 \\ 0 & q_{22} & 0 & \dots & 0 & 0 \\ 0 & 0 & q_{33} & \dots & 0 & 0 \\ & \cdot & \cdot & \cdot & & \\ & & \cdot & \cdot & \cdot & \\ 0 & 0 & 0 & \dots & 0 & q_{nn} \end{pmatrix},$$

and

$$\mathbf{H} = \begin{pmatrix} 0 & q_{21} & q_{31} & \dots & q_{n-1,1} & q_{n1} \\ 0 & 0 & q_{32} & \dots & q_{n-1,2} & q_{n2} \\ & & \cdot & \cdot & \cdot & \\ & & & \cdot & \cdot & \cdot \\ 0 & 0 & 0 & \dots & 0 & q_{n,n-1} \\ 0 & 0 & 0 & \dots & 0 & 0 \end{pmatrix}.$$

Therefore, the system of linear equations $\mathbf{Q}^T\mathbf{y}=\mathbf{0}$ can be written as $(\mathbf{B}+\mathbf{D}+\mathbf{H})\mathbf{y}=\mathbf{0}$. In the Jacobi method, $(\mathbf{B}+\mathbf{D}+\mathbf{H})\mathbf{y}=\mathbf{0}$ is re-written as $\mathbf{D}\mathbf{y}=-(\mathbf{B}+\mathbf{H})\mathbf{y}$, or $\mathbf{y}=-\mathbf{D}^{-1}(\mathbf{B}+\mathbf{H})\mathbf{y}$ from which we construct the iteration

$$\mathbf{y}^{(k)} = -\mathbf{D}^{-1}(\mathbf{B}+\mathbf{H})\mathbf{y}^{(k-1)}.$$

The iteration is stopped when $|\mathbf{y}^{(k)}-\mathbf{y}^{(k-1)}|<\varepsilon$. An alternative iterative method is the power method. In this method, we solve $\mathbf{P}^T\mathbf{y}=\mathbf{y}$ using the iteration equation

$$\mathbf{y}(k) = \mathbf{P}^T\mathbf{y}(k-1).$$

As k tends to infinity, $\mathbf{y}^{(k)}$ approaches $\eta_1\mathbf{y}^{(k-1)}$, where η_1 is the dominant eigenvalue which, in this case, is equal to 1. The power method converges as long as the rate matrix is irreducible and aperiodic. The rate of convergence of the power method depends upon the value of the second dominant eigenvalue. The closer this value is to one, the slower is the rate of convergence.

The Jacobi method usually converges slowly. A modification of it is the well-known Gauss-Seidel method which generally converges faster than the Jacobi method. In the Gauss-Seidel method we re-write $(\mathbf{B}+\mathbf{D}+\mathbf{H})\mathbf{y}=\mathbf{0}$ as $(\mathbf{B}+\mathbf{D})\mathbf{y}=-\mathbf{H}\mathbf{y}$, or $\mathbf{y}=-(\mathbf{B}+\mathbf{D})^{-1}\mathbf{H}\mathbf{y}$ from which we construct the iteration

$$\mathbf{y}^{(k)} = -(\mathbf{B}+\mathbf{D})^{-1}\mathbf{H}\mathbf{y}^{(k-1)}.$$

The Gauss-Seidel method can be modified so that its convergence is further improved. This modification results in the successive overrelaxation method (SOR). In the SOR method, we re-write $\mathbf{Q}^T\mathbf{y}=\mathbf{0}$ as $\omega(\mathbf{B}+\mathbf{D}+\mathbf{H})\mathbf{y}=\mathbf{0}$, were ω is the relaxation parameter. The system $\omega(\mathbf{B}+\mathbf{D}+\mathbf{H})\mathbf{y}=\mathbf{0}$ can be re-written as $\omega(\mathbf{B}+\mathbf{D})\mathbf{y}+\mathbf{D}\mathbf{y}=-\omega\mathbf{H}\mathbf{y}+\mathbf{D}\mathbf{y}$, or $\mathbf{y}=-(\omega\mathbf{B}+\mathbf{D})^{-1}(\omega\mathbf{H}+(\omega-1)\mathbf{D})\mathbf{y}$ which is used to construct the iteration

$$\mathbf{y}^{(k)}=-(\omega\mathbf{B}+\mathbf{D})^{-1}(\omega\mathbf{H}+(\omega-1)\mathbf{D})\mathbf{y}^{(k-1)}.$$

When $\omega=1$, we have $\mathbf{y}^{(k)}=-(\mathbf{B}+\mathbf{D})^{-1}\mathbf{H}\mathbf{y}^{(k-1)}$, which is the Gauss-Seidel method. The SOR method can converge faster than the Gauss-Seidel method by appropriately choosing a value for ω.

2.5. The matrix-geometric procedure

This procedure can be used when the rate matrix has a particular block lower (or upper) Hessenberg structure. A special case of this general structure, which is commonly found in queueing systems, is the case where the rate matrix has the following block tri-diagonal structure

$$\mathbf{Q} = \begin{pmatrix} \mathbf{A}' & \mathbf{B}' & & & \\ \mathbf{C}' & \mathbf{A} & \mathbf{B} & & \\ & \mathbf{C} & \mathbf{A} & \mathbf{B} & \\ & & \mathbf{C} & \mathbf{A} & \mathbf{B} \\ & & & \cdot & \cdot \quad \cdot \end{pmatrix}, \qquad (2.4.1)$$

where each capital letter represents a block sub-matrix. An example of such a rate matrix structure was given in section 2.3 for a two-node queueing network with blocking.

Let $\mathbf{x}$ be the vector of the steady-state probabilities associated with $\mathbf{Q}$, i.e. $\mathbf{xQ}=\mathbf{0}$ and $\mathbf{xe}=1$, where $\mathbf{e}$ is a vector of 1's. We partition $\mathbf{x}$ conformally with the blocks of matrix $\mathbf{Q}$. For instance, in the example given in section 2.3, subvector $\mathbf{x}_0$ has 3 components and subvector $\mathbf{x}_i$, i=1,2,..., has 4 components. $(\mathbf{x}_0,\mathbf{x}_1,...,\mathbf{x}_i,...)\mathbf{Q}=\mathbf{0}$ can then be written out as

$$\mathbf{x}_0\mathbf{A}' + \mathbf{x}_1\mathbf{C}' = \mathbf{0}$$

$$\mathbf{x}_0\mathbf{B}' + \mathbf{x}_1\mathbf{A} + \mathbf{x}_2\mathbf{C} = \mathbf{0}$$

$$\mathbf{x}_i\mathbf{B} + \mathbf{x}_{i+1}\mathbf{A} + \mathbf{x}_{i+2}\mathbf{C} = \mathbf{0}, \; i=1,2,....$$

By manipulating the above equations it can be shown that $\mathbf{x}_i$, i=1,2,..., can be expressed in terms of $\mathbf{x}_1$ as follows:

$$\mathbf{x}_i = \mathbf{x}_1\mathbf{R}^{i-1}, \; i=1,2,..., \qquad (2.4.2)$$

where $\mathbf{R}$ is the minimal non-negative solution of the matrix equation

$$\mathbf{B} + \mathbf{RA} + \mathbf{R}^2\mathbf{C} = \mathbf{0}. \qquad (2.4.3)$$

In certain cases, it is possible to obtain a closed-form expression for $\mathbf{R}$. In general, however, $\mathbf{R}$ is obtained recursively by rewriting (2.4.3) as

$$\mathbf{R}_{n+1} = -\mathbf{B}(\mathbf{A}^{-1}) - (\mathbf{R}_n^2\mathbf{C})(\mathbf{A}^{-1}) \qquad (2.4.4)$$

and setting $\mathbf{R}_0=\mathbf{0}$. Subvectors $\mathbf{x}_0$ and $\mathbf{x}_1$ can be obtained from the equations

$$\begin{cases} \mathbf{x}_0\mathbf{A}' + \mathbf{x}_1\mathbf{C}' = \mathbf{0} \\ \mathbf{x}_0\mathbf{B}' + \mathbf{x}_1(\mathbf{A}+\mathbf{R}\mathbf{C}) = \mathbf{0} \end{cases} \tag{2.4.5}$$

and the normalizing condition. In the numerical procedures described in section 2.4, the normalization of the stationary probability vector was handled by replacing one of the equations by the normalizing equation. In the matrix-geometric procedure, the normalizing equation can be used in conjunction with (2.4.5) in order to solve for $\mathbf{x}_0$ and $\mathbf{x}_1$. Alternatively, it can be handled as follows. One of the elements of $\mathbf{x}_0$ or $\mathbf{x}_1$ is assigned the value 1. Then, all the other probabilities are calculated accordingly using (2.4.5) and (2.4.2). Because of this, it is possible that some probabilities may have a value greater than 1. (We may not even call them probabilities at this stage!). Finally, each of these probabilities is divided by the sum of all the probabilities so that they all add up to 1. Attention should be paid to the value assigned to the first probability. It is common practice to assign the value of 1. However, depending on the size of the state space, assigning the value of 1 may cause the sum of all probabilities to overflow. This overflow problem may be avoided by assigning a value smaller than 1. In certain cases, truncation of the state space may be necessary.

Below we give three examples of the matrix-geometric procedure. In the first example, we analyze a single server queue with a phase-type service time. The capacity of the queue is assumed to be infinite. In the second example, we analyze the same queue only we assume that it has a finite capacity. In these two cases, we can obtain a closed-form expression for matrix $\mathbf{R}$. Finally, in the third example, we analyze a three-node queueing network with blocking.

Matrix-geometric solution of the M/PH/1 queue

Let us consider an infinite capacity queue served by a single server with Poisson arrivals and a phase-type service time. Such a queueing system is notated as M/PH/1. Let λ be the arrival rate, and let the phase-type service be described by $(\boldsymbol{\alpha},\mathbf{T})$. The mean service time is $\mu^{-1}=-\boldsymbol{\alpha}\mathbf{T}^{-1}\mathbf{e}$. We assume that the node is stable, i.e. $\rho=\lambda/\mu<1$. Now, let us define the Markov the Markov process $\{X_t, Y_t\}$, $t \geq 0$, where X_t represents the number of customers in the system (including the one in service) at time t, and $Y_t = 1,2,...,k$, represents the phase of the service process at time t. The state space of this system is $E=\{0, (i,j);\ i=1,2,...,\ j=1,2,...,k\}$. These states can be ordered lexicographically, and then organized into *groups* or *levels* as shown in table 2.1. The organization of states is such that all the states which have the same number of customers in the system are grouped together. This grouping is used below to describe the block structure of the rate matrix.

Let us now consider an example of an M/PH/1 node as shown in figure 2.2. As above, let (i,j) indicate the state of the system, where i, i=0,1,..., is the number of customers in the node and j, j=1,2,3, is the phase of the service. The states of the system are ordered lexicographically and then grouped into levels as described above. The phase-type service process of this queue is defined by $(\boldsymbol{\alpha}, \mathbf{T})$, where $\boldsymbol{\alpha}=(1,0,0)$, and

$$\mathbf{T} = \begin{pmatrix} -\mu_1 & \mu_1 a_{12} & \mu_1 a_{13} \\ \mu_2 a_{21} & -\mu_2 & \mu_2 a_{23} \\ \mu_3 a_{31} & 0 & -\mu_3 \end{pmatrix}, \ \mathbf{T}^{o} = \begin{pmatrix} 0 \\ \mu_2 a_{24} \\ \mu_3 a_{34} \end{pmatrix}.$$

The rate matrix **Q** of this M/PH/1 queue is

Table 2.1: States and associated levels

State	Level
0	0
(1,1),(1,2),...,(1,k)	1
(2,1),(2,2),...,(2,k)	2
.	
.	
.	
(i,1),(i,2),...,(i,k)	i
.	
.	
.	

$$
\mathbf{Q}=\left(\begin{array}{cccccccccc}
* & \lambda & 0 & 0 & & & & & & \\
0 & * & \mu_1\alpha_{12} & \mu_1\alpha_{13} & \lambda & & & & & \\
\mu_2\alpha_{24} & \mu_2\alpha_{21} & * & \mu_2\alpha_{23} & & \lambda & & & & \\
\mu_3\alpha_{34} & \mu_3\alpha_{31} & 0 & * & & & \lambda & & & \\
 & 0 & 0 & 0 & * & \mu_1\alpha_{12} & \mu_1\alpha_{13} & \lambda & & \\
 & \mu_2\alpha_{24} & 0 & 0 & \mu_2\alpha_{21} & * & \mu_2\alpha_{23} & & \lambda & \\
 & \mu_3\alpha_{34} & 0 & 0 & \mu_3\alpha_{31} & 0 & * & & & \lambda \\
 & & & & & \cdot & \cdot & \cdot & \cdot & \\
 & & & & & & \cdot & \cdot & \cdot & \cdot
\end{array}\right)
$$

In the above matrix, row 1 gives the rate of transition from state (0,0) to state (1,1), row 2 gives the rate of transition from state (1,1) to states (1,2), (1,3), (2,1), and so on. Also, each diagonal element marked by a * is the negative sum of the elements in the same row. Following the grouping of states into levels described above, **Q** can be written in the block tri-diagonal form

$$
\mathbf{Q}=\left(\begin{array}{cccccc}
-\lambda & \lambda\boldsymbol{\alpha} & & & & \\
\mathbf{T^o} & \mathbf{T}-\lambda\mathbf{I} & \lambda\mathbf{I} & & & \\
 & \mathbf{T^o}\boldsymbol{\alpha} & \mathbf{T}-\lambda\mathbf{I} & \lambda\mathbf{I} & & \\
 & & \mathbf{T^o}\boldsymbol{\alpha} & \mathbf{T}-\lambda\mathbf{I} & \lambda\mathbf{I} & \\
 & & & \cdot & \cdot & \cdot
\end{array}\right).
$$

In the first row, the transition from level 0 to level 1, i.e. to states (1,1), (1,2), (1,3), indicates the arrival of a customer to an empty queue. The customer starts its service at the first phase with probability 1. The second row gives the transitions from level 1 to levels 0, 1 and 2. A transition to level 0, 1 and 2 indicates a departure from the queue, a change of phase in the service process, and an arrival to the queue respectively. Transitions

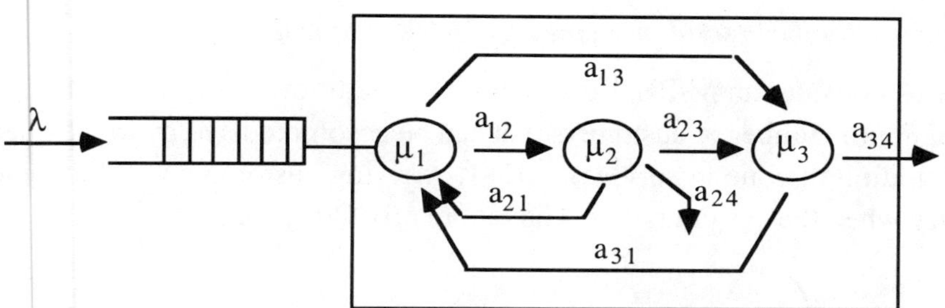

Figure 2.2: An M/PH/1 queue.

from the other levels can be similarly interpreted. We note that the rate matrix of any M/PH/1 queue has the same structure as above, where $\boldsymbol{\alpha}$ and $\mathbf{T}$ are defined from the phase-type service distribution.

Let $\mathbf{x}$ be the stationary probability vector of an M/PH/1 queue such that $\mathbf{xQ=0}$. We partition $\mathbf{x}$ so that each subvector corresponds to a level, i.e. $\mathbf{x}=(x_0,\mathbf{x}_1,\ldots,\mathbf{x}_i,\ldots)$, where $\mathbf{x}_i=(x_{i1},x_{i2},\ldots,x_{ik})$, i=1,2,..., and x_{ij} is the steady-state probability that the system is in state (i,j). Then, $\mathbf{xQ=0}$ can be written out as

$$-\lambda x_0 + \mathbf{x}_1\mathbf{T}^o = 0$$

$$\lambda \mathbf{x}_0\boldsymbol{\alpha} + \mathbf{x}_1(\mathbf{T}-\lambda\mathbf{I}) + \mathbf{x}_2\mathbf{T}^o\boldsymbol{\alpha} = \mathbf{0}$$

$$\vdots$$

$$\lambda \mathbf{x}_{i-1}\mathbf{I} + \mathbf{x}_i(\mathbf{T}-\lambda\mathbf{I}) + \mathbf{x}_{i+1}\mathbf{T}^o\boldsymbol{\alpha} = \mathbf{0},\ i\geq 2.$$

By manipulating these equations, it can be shown that

$$\mathbf{x}_1 = x_0\boldsymbol{\alpha}\mathbf{R}$$

$$\mathbf{x}_i = \mathbf{x}_{i-1}\mathbf{R},\ i=2,3,\ldots,$$

where

$$\mathbf{R} = \lambda\ (\lambda\mathbf{I}-\lambda\mathbf{e}\boldsymbol{\alpha}-\mathbf{T})^{-1}$$

and $x_0=1-\lambda\mu^{-1}=1-\lambda\boldsymbol{\alpha}\mathbf{T}^{-1}\mathbf{e}$. Thus, $\mathbf{R}$ is obtained by inverting $\lambda\mathbf{I}-\lambda\mathbf{e}\boldsymbol{\alpha}-\mathbf{T}$, and subsequently the stationary probability vector is calculated one subvector at a time.

Matrix-geometric solution of the M/PH/1/K+1 queue

Let us consider an M/PH/1 queue with the additional assumption that the maximum number of customers that can be accommodated by in the queue (including the one in service) is K+1. Arriving customers will be denied entry when the system is full. The rate matrix $\mathbf{Q}$ is given by

$$\mathbf{Q} = \begin{pmatrix} -\lambda & \lambda\alpha & & & & & \\ \mathbf{T}^{o} & \mathbf{T}\text{-}\lambda\mathbf{I} & \lambda\mathbf{I} & & & & \\ & \mathbf{T}^{o}\alpha & \mathbf{T}\text{-}\lambda\mathbf{I} & \lambda\mathbf{I} & & & \\ & & & \cdot & \cdot & \cdot & \\ & & & & \mathbf{T}^{o}\alpha & \mathbf{T}\text{-}\lambda\mathbf{I} & \lambda\mathbf{I} \\ & & & & & \mathbf{T}^{o}\alpha & \mathbf{T} \end{pmatrix}$$

Let $\mathbf{x}=(x_0,\mathbf{x}_1,...,\mathbf{x}_{K+1})$ be the stationary vector such that $\mathbf{xQ=0}$. Then, following similar arguments as in the case of the infinite capacity M/PH/1 queue, we obtain

$$\mathbf{x}_i = x_0\alpha\mathbf{R}^i \ , \ i=1,2,...,K,$$

$$\mathbf{x}_{K+1} = x_0\alpha\mathbf{R}^K(-\lambda\mathbf{T}^{-1}),$$

where $\mathbf{R}$ is given by

$$\mathbf{R} = \lambda(\lambda\mathbf{I}-\lambda\mathbf{e}\alpha-\mathbf{T})^{-1}$$

and

$$x_0 = \{\alpha\,[\sum_{i=0}^{K}\mathbf{R}^i - \lambda\mathbf{R}^K\mathbf{T}^{-1}]\mathbf{e}\}^{-1}.$$

Matrix-geometric solution of a queueing network with blocking

Let us consider the three-node open queueing network shown in figure 2.3. Each node is represented by a single server queue. Let m_i and μ_i be the capacity and the service rate of node i respectively. The capacity of the first node is infinite, and the capacity of the remaining nodes is finite. External arrivals occur only to node 1 at the rate λ. All service times and external inter-arrival times are exponentially distributed. Customers are served in a FIFO manner.

The state of this queueing network can be described by the vector (n_1,n_2,n_3), where $n_1=0,1,\ldots$, $n_2=0,1,\ldots,m_2+1$, and $n_3=0,1,\ldots,m_3+4$. The random variable n_i, $i=1,2,3$, indicates the number of customers in node i; it varies from 0 to m_i for nodes 2 and 3, and it takes any non-negative integer number for node 1. The additional values associated with random variables n_2 and n_3 are used to describe the various blocking states. In particular, $n_2=m_2+1$ denotes that node 1 is blocked by node 2, $n_3=m_3+1$ denotes that node 1 is blocked by node 3, $n_3=m_3+2$ denotes that node 2 is blocked by node 3, $n_3=m_3+3$ denotes that nodes 1 and 2 are blocked by node 3 in that order, and $n_3=m_3+4$ denotes that nodes 2 and 1 are blocked by node 3 in that order. The states can be easily generated in a lexicographical order using the following algorithm

```
ind = 1
for i = 0 to m1 do
    for j = 0 to m2+1 do
        for k = 0 to m3+4 do
            if (i,j,k) is feasible then
                begin
                    states(ind) = (i,j,k)
                    ind = ind + 1
                end;
```

Eventually, array *states* will contain all feasible states. This algorithm simple generates each possible combination of the variables n_1,n_2 and n_3, and then tests whether it is feasible or not. (When dealing with more complex queueing systems it is preferable to have an algorithm which will

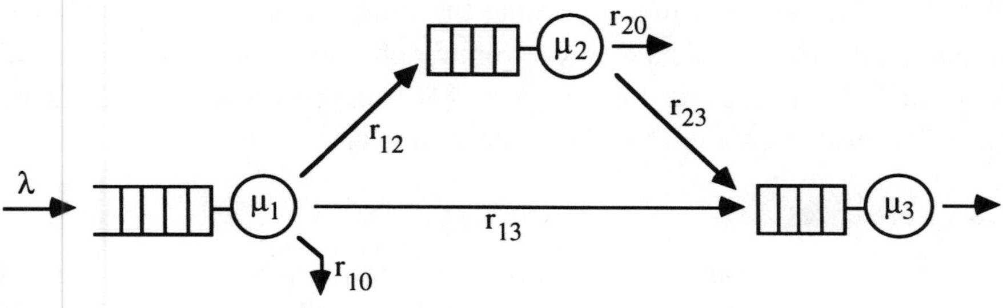

Figure 2.3: A three-node feed-forward configuration.

generate only the feasible states.) The generation of the rate matrix can now be accomplished by going through the list of states and generating all the feasible transitions out of each state and the associated rate of transition. The resulting rate matrix **Q** has the block tri-diagonal structure given by (2.4.1). Due to the repetitive nature of the rate matrix, it suffices to generate sub-matrices **A′**, **B′**, **C′**, **A**, **B** and **C**.

Once the necessary submatrices have been set-up, the stationary probability vector can be obtained by first solving recursively for **R** using (2.4.4), and subsequently calculating $\mathbf{x}_i$, i=0,1,..., from (2.4.2) and (2.4.5). The stability condition of the queueing network can be obtained from the matrix-geometric procedure. The general form of this condition is given by

$$\mathbf{zCe} > \mathbf{zBe}$$

where **z** is obtained solving **z(C+A+B)=0**.

The order M of submatrices **A**, **B** and **C** is

$$M = (m_2+1)(m_3+1) + 4m_2 + m_3 + 3.$$

The order of matrix **R** is also M. For the three-node queueing network the matrix-geometric procedure does not require much memory even when m_2 and m_3 are large. The memory requirements, however, drastically increase as the number of nodes in the queueing network increases. Also, it has been observed that the computation of the steady-state probabilities is the major contributor to the time complexity of the matrix-geometric procedure. That is, the solution for **R** and $\mathbf{x}_i$, i=0,1,..., require much more time than the generation of the states and the rate matrix. Furthermore, the computation of **R** is the most time consuming operation. The convergence of R depends on the number of nodes, the topology of the network, the buffer sizes, and the congestion level in the network. The first three factors basically determine the order of **R**. As the order of R increases, the time it takes **R** to converge increases as well. A non-trivial factor that also affects the convergence of **R** is the congestion level. As the traffic intensity increases the convergence time also increases.

2.6. References

The analysis of nearly completely decomposable Markov processes referred to in the introductory section is discussed in detail in Courtois

(1977) See also Courtois and Semal (1986). The material in section 2.2. was presented following Stewart (1988). For further details on the numerical methods presented in section 2.4 see Burden and Faires (1989). Also, the interested reader can find applications of Gaussian elimination to queueing models in Stewart (1989). The matrix-geometric procedure presented in section 2.5 is due to Neuts (1981). The memory and cpu complexity of the matrix-geometric procedure when used to analyze open queueing networks with blocking was examined by Onvural, Perros, and Altiok (1987). Latouche and Ramaswami (1992) proposed an algorithm which drastically reduces the number of iterations required to calculate matrix **R** for quasi birth-death processes. Finally, we note that there are several software packages available which facilitate the generation of Markov chains and compute the transient probability distribution and the stationary probability vector, such as MARCA (see Stewart (1991)).

Bibliography

Burden, R. L. and Faires, J. D., *Numerical Analysis*, (PWS-Kent, 1989)

Courtois, P.-J., *Decomposability, Queueing and Computer System Applications,* (Academic Press,1977).

Courtois, P.-J. and Semal, P., Computable bounds for conditional steady-state probabilities in large Markov chains and queueing models, *IEEE JSAC* **4,6** (1986) 920-936.

Latouche, G. and Ramaswami, Y., *A logarithmic reduction algorithm for quasi birth-death processes*, Belcore Tech. Rept. TM-TFV-021374, 1992.

Neuts, M. F., *Matrix-geometric solutions in stochastic models - an algorithmic approach*, (The John Hopkins Univ. Press., 1981).

Onvural, R. O., Perros, H. G. and Altiok, T., On the complexity of the matrix-geometric solution of exponential open queueing networks with blocking, in: Pujolle, Fdida, and Horlait, (Eds.), *Proc. Int. Workshop on Modelling Techniques and Performance Evaluation*, (North-Holland, Amsterdam, 1987) 3-12.

Stewart, W. J., *Markov chains and stochastic matrices*, Tech. Rept., Computer Science Dept., North Carolina State University, 1988.

Stewart, W. J., Recursive procedures for the numerical solution of Markov chains, in: Perros and Altiok, (Eds.), *Proc. First International Workshop on Queueing networks with blocking*, (North-Holland, Amsterdam, 1989) 229-247.

Stewart, W. J., MARCA: Markov chain analyzer, a software package for Markovian modelling, in: Stewart, (Ed.), *Numerical solutions of Markov chains*, (Dekker, 1991) 37-61.

CHAPTER 3

Two-Node Open Queueing Networks With Blocking

3.1. Introduction

In this Chapter, we study various two-node open queueing network with blocking. In section 3.2, we consider a two-node queueing network with blocking with exponentially distributed service times. We obtain the condition for stability and we also derive closed form solutions for various special cases. In section 3.3, we examine the case of unreliable servers assuming exponential service times and also constant synchronized service times. In section 3.4, we consider a two-node queueing network with blocking assuming that the first node is saturated and that the service times have a phase-type distribution. We present two methods for the analysis of this queueing network, namely, a matrix-recursive method and a matrix-geometric method. Finally, relevant references are given in section 3.5.

3.2. Exponential service times

Let us consider two nodes in tandem as shown in figure 3.1. The first node has an infinite capacity and the second node has a finite capacity of size m. Service at each node is provided by a single exponential server. External arrivals join the first node in a Poisson fashion at the rate λ. Let μ_1 and μ_2 be the service rate at the first and second server respectively. Customers in each node are served in a FIFO manner. We assume blocking-after-service. That is, the first server becomes blocked when a customer completes its service at a time the second node is full. The server remains blocked and it cannot serve any other customer in its queue until a departure occurs from node 2.

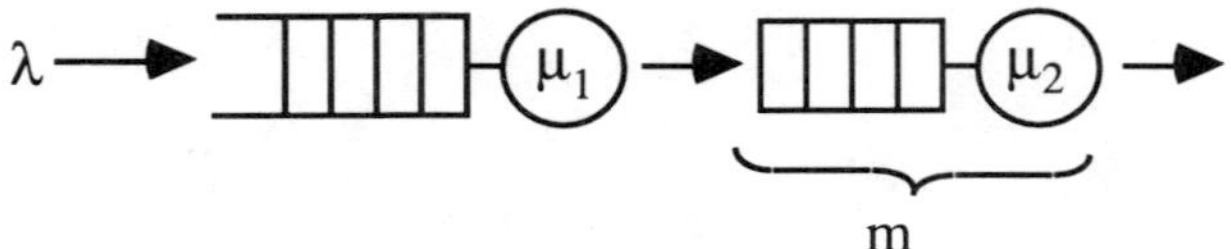

Figure 3.1: A two-node open queueing network with blocking.

The state of the above queueing network can be described by the random variables (n_1,n_2), where n_1 indicates the number of customers in the first node and n_2 indicates the number of customers in the second node. We have $n_1=0,1,...,$ and $n_2=0,1,...,m,m+1$. $n_2=m+1$ indicates that the second node is full and the first server is blocked. Thus, any state $(i,m+1)$, $i=1,2,...,$ indicates that the first server is blocked and that there are i customers in the first node. Of these i customers, one has received its service but it is blocked from entering node 2 and the remaining i-1 are waiting to be served.

For presentation purposes we will consider the case of m=2. As it will be seen below, this is not a trivial case. Let $p(n_1,n_2)$ be the steady-state probability that the system is in state (n_1,n_2). The steady-state equations are

$$\lambda p(0,0) = \mu_2 p(0,1)$$

$$(\lambda+\mu_1)p(n_1,0) = \mu_2 p(n_1,1) + \lambda p(n_1-1,0), \quad n_1=1,2,... .$$

$$(\lambda+\mu_2)p(0,1) = \mu_1 p(1,0) + \mu_2 p(0,2)$$

$$(\lambda+\mu_1+\mu_2)p(n_1,1) = \mu_1 p(n_1+1,0) + \mu_2 p(n_1,2) + \lambda p(n_1-1,1), \quad n_1=1,2,... .$$

$$(\lambda+\mu_2)p(0,2) = \mu_1 p(1,1) + \mu_2 p(1,3)$$

$$(\lambda+\mu_1+\mu_2)p(n_1,2) = \mu_1 p(n_1+1,1)+\mu_2 p(n_1+1,3)+\lambda p(n_1-1,2), \quad n_1=1,2,... .$$

$$(\lambda+\mu_2)p(1,3) = \mu_1 p(1,2)$$

$$(\lambda+\mu_2)p(n_1,3) = \mu_1 p(n_1,2) + \lambda p(n_1-1,3), \quad n_1=2,3,... .$$

We also have the normalizing equation $\sum p(n_1,n_2)=1$. Now, define the generating functions

$$g_k(z) = \sum_{n_1=0}^{\infty} p(n_1,k)z^{n_1}, \quad k=0,1,2,$$

$$g_3(z) = \sum_{n_1=1}^{\infty} p(n_1,3)z^{n_1-1}.$$

Then, the above set of steady-state equations can be written as

$$g_0(z)\,[\lambda(1-z)+\mu_1] = \mu_2 g_1(z) + \mu_1 p(0,0)$$

$$g_1(z)[\lambda(1-z)+\mu_1+\mu_2] = \frac{\mu_1}{z} g_0(z) + \mu_2 g_2(z) + \mu_1 p(0,1) - \frac{\mu_1}{z} p(0,0)$$

$$g_2(z)[\lambda(1-z)+\mu_1+\mu_2] = \frac{\mu_1}{z} g_1(z) + \mu_2 g_3(z) + \mu_1 p(0,2) - \frac{\mu_1}{z} p(0,1)$$

$$g_3(z)[\lambda(1-z)+\mu_2] = \frac{\mu_1}{z} g_2(z) - \frac{\mu_1}{z} p(0,2)\ .$$

The normalizing equation becomes

$$\sum_{k=1}^{3} g_k(1) = 1.$$

The total number of independent equations is six, namely the above four equations, the normalizing condition, and equation $\lambda p(0,0)=\mu_2 p(0,1)$ from the steady-state equations. However, the total number of unknowns is seven, namely, $g_0(z)$, $g_1(z)$, $g_2(z)$, $g_3(z)$, $p(0,0)$, $p(0,1)$ and $p(0,2)$. Thus, we are short by one equation. In the case of general m, the number of unknowns is 2m+3. This consists of m+2 unknown generating functions and m+1 unknown probabilities $p(0,0)$, $p(0,1)$,...,$p(0,m)$. The number of available independent equations is m+4, thus we are m-1 equations short. In view of this, it is not possible to solve the above system of equations for $g_k(z)$, k=1,2,...,m+1, except when m=1.

One way to obtain the additional equation(s) required for the solution of the above system is to exploit the fact that a generating function is analytic within the unit circle. In particular, let us consider a generating function

$g(z)$ and let us assume that it can be written as a ratio of two polynomials, i.e. $g(z)=f_1(z)/f_2(z)$. Now, in order for $g(z)$ to be analytic within the unit circle, all zeroes of $f_2(z)$ within the unit circle have to be zeroes of $f_1(z)$ as well. Let the zeroes of $f_2(z)$ within the unit circle be $\zeta_1,\zeta_2,...,\zeta_k$. Then, we have that $f_1(\zeta_i)=0$, $i=1,2,...,k$, which gives us k additional equations. This argument has been used successfully to analyze different queueing systems. In general, the zeroes are calculated numerically, though in several cases, one can obtain their exact closed-form expression.

Below, we present various closed-form results related to the queueing network described above. Due to the equivalencies between the blocking mechanisms: blocking-after-service, blocking-before-service and repetitive service, we will only consider the queueing network under blocking-after-service.

Two limiting cases

In this section, we examine the queueing network described above when $\mu_1\to\infty$ and when the first node is saturated.

a. $\mu_1 \to \infty$

In this case, a customer at the first node receives an infinitesimal amount of service. If the customer arrives at the queueing network when the second node contains less than m customers, the customer goes through the first node and it immediately joins the second node. If the second node contains m customers, the arriving customer is immediately blocked and the first server is blocked as well. When a customer departs from the second node, the first server becomes unblocked for an infinitely small amount of time and then it gets blocked again, if there are customers waiting in the first node. The queueing network, therefore, is reduced to an M/M/1 queue with traffic intensity $\rho=\lambda/\mu_2$. In view of this, the steady-state probability that there are n, $n=0,1,...$, customers in the queueing network is $p(n)=(1-\rho)\rho^n$. The probability that there are n_2 customers in the second node is simply

$$p_2(n_2) = (1-\rho)\rho^{n_2},\ n_2=0,1,...,m-1,$$

$$p_2(m) = \sum_{n\geq m}(1-\rho)\rho^n=\rho^m.$$

The probability that there are n_1, n_1=1,2,.., customers in the first node is $p_1(n_1)=(1-\rho)\rho^{m+n_1}$, and

$$p_1(0) = \sum_{n=0}^{m}(1-\rho)\rho^n = 1-\rho^{m+1}.$$

b. Saturated first node

Let us now consider the queueing network when the first node is *saturated*. We say that a node is saturated when there is always at least one customer waiting for service, i.e the node is never empty. This case is often described in production systems as a server with an unlimited supply of raw material. For given λ, μ_2 and m, let μ_1^0 be the service rate at the first node below which the first node gets saturated. Then, when $\mu_1 \leq \mu_1^0$ the first server is either busy serving or blocked. A customer upon completion of its service at the first server gets blocked if at that moment the second node is full. The blocking customer remains in front of the first server until a departure occurs from the second node. At that instant it moves to the mth position of the second node and the first server becomes unblocked. During the blocking period, therefore, the first server can be seen as providing an additional storage space to the second node. Furthermore, during the blocking period the first server does not serve any other customers. That is, no more arrivals occur at the second node. Due to the exponential assumption, this is equivalent to saying that arrivals do occur at the second node at the rate μ_1 but they are lost. In view of this, the second node becomes an M/M/1/m+1 queue with an overall arrival rate μ_1 and service rate μ_2. Hence, the probability that there are n_2 customers in the second node is $p_2(n_2) = (1-\sigma)\sigma^{n_2}/(1-\sigma^{m+1})$, where $\sigma=\mu_1/\mu_2$.

This limiting case has been used extensively to model two-stage production systems. The quantity μ_1^0 is obtained below.

The blocking probability

The blocking probability is defined as the probability that a customer upon service completion at the first server will get blocked. That is, at that instant the second node is full. This probability (call it π) has been used in several approximation algorithms (see Chapters 4 and 5). Below, we obtain the exact expression for this probability when $\mu_1 \to \infty$ (call it π^∞) and when $\mu_1 \leq \mu_1^0$ (call it π^0).

When $\mu_1 \to \infty$, the queueing network reduces to an M/M/1 queue as discussed above. A customer will get blocked with probability 1 if upon arrival it sees m or more customers in the M/M/1 queue. A customer will not get blocked if upon arrival it sees less than m customers. Thus, π^∞=1xProb(an arrival sees m or more customers). Due to the PASTA argument (i.e. Poisson arrivals see time averages), we have

$$\pi^\infty = \sum_{n \geq m} p(n) = (1-\rho) \sum_{n \geq m} \rho^n = \rho^m , \tag{3.2.1}$$

where $\rho = \lambda/\mu_2$. We note that π^∞ coincides with the time average probability that there are m customers in the second node.

Now, let us consider the case where $\mu_1 \leq \mu_1^0$. As it was mentioned above, the second node behaves as an M/M/1/m+1 queue. The effective arrival rate $\bar{\lambda}$ at the second node is

$$\bar{\lambda} = \mu_1[1-p_2(m+1)]$$

where

$$p_2(m+1) = \frac{(1-\sigma)\sigma^{m+1}}{1-\sigma^{m+2}} ,$$

σ being equal to μ_1/μ_2. Applying Little's relation to the (m+1)st position we obtain

$$\bar{\lambda}\pi^0/\mu_2 = p_2(m+1),$$

where $\bar{\lambda}\pi^0$ is the rate at which customers enter the (m+1)st position, $1/\mu_2$ is the mean time a customer spends in this position, and $p_2(m+1)$ is the mean number of customers occupying this position. Combining the above two expressions we have

$$\mu_1[1-p_2(m+1)]\pi^0/\mu_2 = p_2(m+1)$$

or

$$\sigma\left(1 - \frac{(1-\sigma)\sigma^{m+1}}{1-\sigma^{m+2}}\right)\pi^0 = \frac{(1-\sigma)\sigma^{m+1}}{1-\sigma^{m+2}}$$

or

$$\pi^0 = \frac{(1-\sigma)\sigma^m}{1-\sigma^{m+1}} = \frac{\sigma^m}{1+\sigma+...+\sigma^m}. \tag{3.2.2}$$

We note that π^0 coincides with the time average probability that the second node is full when it is analyzed as an M/M/1/m queue with traffic intensity $\sigma=\mu_1/\mu_2$. The quantity $1-\pi^0$ can be seen as the percentage of time that the first server is busy serving. For the special case $\mu_1=\mu_2$, we have that $1-\pi^0= m/(m+1)$.

Condition for stability

For given λ, μ_2 and m, the quantity μ_1^0 was defined as the critical service rate of the first server below which the first queue becomes unstable. If the first node is stable, then the effective rate into the second node is λ. However, if $\mu_1 \leq \mu_1^0$, then $\bar{\lambda}=\mu_1[1-p_2(m+1)]$. Now, for $\mu_1=\mu_1^0$, we have that $\lambda=\bar{\lambda}$, or

$$\lambda = \mu_1^0\left(1 - \frac{(1-\sigma_0)\sigma_0^{m+1}}{1-\sigma_0^{m+2}}\right), \tag{3.2.3}$$

where $\sigma_0=\mu_1^0/\mu_2$. The quantity μ_1^0 can be computed numerically using the above expression.

The condition for stability of the queueing network is simply the condition for stability of the first node. We first observe that when the first node is saturated, the maximum departure rate from this node is $\mu_1[1-p_2(m+1)]$, seeing that $1-p_2(m+1)$ is the percentage of time that the first server is not blocked. Hence, in order for the first server to be stable we should have $\mu_1[1-p(m+1)]>\lambda$, or

$$\lambda < \mu_1\left(1 - \frac{(1-\sigma)\sigma^{m+1}}{1-\sigma^{m+2}}\right)$$

or

$$\lambda/\mu_1 < \frac{1+\sigma+...+\sigma^m}{1+\sigma+...+\sigma^{m+1}}. \tag{3.2.4}$$

Multiple servers

We now consider a two-node queueing network with blocking assuming that each queue is served by multiple servers as shown in figure 3.2. Let k_1 and k_2 be the number of servers at the first and second node respectively. Customers arrive at the system in a Poisson fashion at the rate λ, and the service times at the first and second node are exponentially distributed with a mean $1/\mu_1$ and $1/\mu_2$ respectively. Finally, let m, $m \geq k_2$, be the maximum capacity of the second node. That is, there can be k_2 customers in service and $m-k_2$ customers waiting for service.

Upon completion of service at the first node, a customer will get blocked if at that moment the second node is full. Its server will also get blocked. When a departure occurs from the second node, one of the blocked customers will enter the second node and its associated server will become unblocked.

We now proceed to obtain a closed-form solution for the queue-length distribution of this queueing network assuming that the first node is saturated. (We note that this model is equivalent to an $M/M/m+k_1$ queue with state-dependent arrivals.) Let p(n), $n=0,1,\ldots,m,m+1,\ldots,m+k_1$, be the probability that there are n customers in the second node including the customers in the first node that are blocked. The steady-state equations associated with p(n) are

$$k_1\mu_1 p(0) = \mu_2 p(1)$$

$$(k_1\mu_1+n\mu_2)p(n) = (n+1)\mu_2 p(n+1)+k_1\mu_1 p(n-1), \quad n=1,2,\ldots,k_2-1,$$

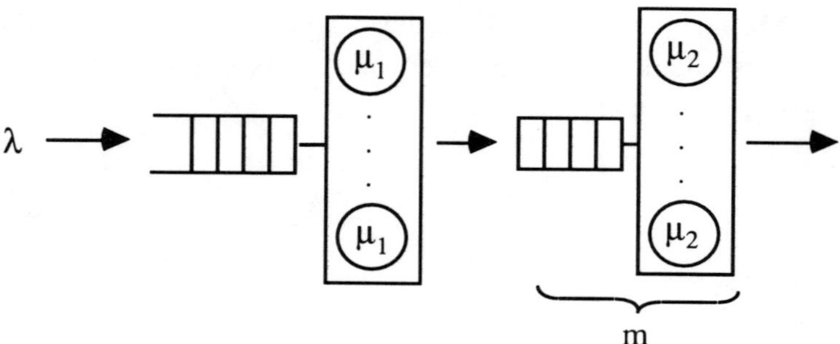

Figure 3.2: Two-node queueing network with multiple servers.

$$(k_1\mu_1+k_2\mu_2)p(n) = k_2\mu_2 p(n+1)+k_1\mu_1 p(n-1),\ n=k_2,\dots,m,$$

$$[(k_1+m-n)\mu_1+k_2\mu_2]p(n) = k_2\mu_2 p(n+1)+[k_1+m-(n-1)]\mu_1 p(n-1),\quad n=m+1,\dots,m+k_1-1,$$

$$k_2\mu_2 p(m+k_1) = \mu_1 p(m+k_1-1).$$

Using the above equations we can obtain after some calculations

$$p(n) = \begin{cases} \dfrac{k_1^n}{n!}\left(\dfrac{\mu_1}{\mu_2}\right)^n p(0) & n=1,\dots,k_2 \\[2ex] \dfrac{k_1^n}{k_2!k_2^{n-k_2}}\left(\dfrac{\mu_1}{\mu_2}\right)^n p(0) & n=k_2+1,\dots,m, \\[2ex] \dfrac{k_1^m\,(k_1)_{(n-m)}}{k_2!\,k_2^{n-k_2}}\left(\dfrac{\mu_1}{\mu_2}\right)^n p(0) & n=m+1,\dots,m+k_1 \end{cases} \tag{3.2.5}$$

where $(k)_i=k(k-1)(k-2)\dots(k-(i-1))$. Probability $p(0)$ is obtained from the normalizing condition.

We note that the departure rate from the second node is

$$\sum_{n=1}^{k_2-1} n\mu_1 p(n) + k_2\mu_2 \sum_{n=k_2}^{m+k_1} p(n), \tag{3.2.6}$$

which is equal to the departure rate from the first node

$$\sum_{n=0}^{m} k_1\mu_1 p(n) + \sum_{n=1}^{k_1-1}(k_1-n)\mu_1 p(m+n), \tag{3.2.7}$$

where $p(n)$, $n=0,1,\dots,m_1+k_1$, is given by (3.2.5). We can now use expression (3.2.6), or (3.2.7), to obtain the condition for stability of this queueing network when the first queue is stable. In this case, expression (3.2.6) gives the maximum departure rate from the queueing network. Therefore, in order for the system to be stable we should have

$$\lambda < \sum_{n=1}^{k_2-1} n\mu_1 p(n) + k_2\mu_2 \sum_{n=k_2}^{m+k_1} p(n).$$

3.3. Unreliable servers

In the previous sections, it was implicitly assumed that a server never fails. In this section, we investigate the case where a server may fail. A failed server becomes operational again after it has been repaired. Below, we examine the case of unreliable servers assuming a) exponentially distributed service times, and b) constant synchronized service times.

Exponential service times

Let us consider the two-node open queueing network shown in figure 3.1. The first node has an infinite capacity and the second node has a finite capacity of size m. Service at each node is provided by a single exponential server, and external arrivals join the first node in a Poisson fashion. We assume blocking-after-service. A server can fail only during the time that it is busy serving. It cannot fail during the time it is idle or blocked. A customer in service is not lost if during its service the server fails. The failure and repair times are exponentially distributed, and blocking-before-service is assumed. Let the vector (n_1,s_1,n_2,s_2) describe the state of the system, where n_1 and n_2 indicate the number of customers in the first and second node respectively, and s_1,s_2 indicate the status of the first and second server respectively. The quantity s_i, i=1,2, takes the values 1 if server i is operational or 0 if the server is broken down. It can be easily verified that if we order the states lexicographically, the non-zero elements of the resulting rate matrix have a block tri-diagonal structure. In view of this, the queueing network can be easily analyzed numerically (see Chapter 2).

Constant synchronized service times: saturated first node

Let us consider the queueing network shown in figure 3.1 under the assumption that both servers begin and end service at the same time. The duration of a service is assumed to be constant and it is taken as the unit time. At the end of a unit time, an operational server may break down with some fixed probability. A server that is broken down at the beginning of a unit time may be repaired at the end of the unit time with some fixed probability. Thus, the only source of randomness in this model is due to the unreliability of the servers. This type of model has been motivated from studies of synchronized assembly machines. We examine this model assuming that there is an unlimited supply of customers in the first node (i.e. the first queue is always saturated). We make the following assumptions.

a) If server i, i=1,2, is operational at the beginning of a unit time, then at the end of the same unit time it may break down with probability p_i, or it may remain operational with probability $\bar{p}_i= 1-p_i$.
b) If server i, i=1,2, is broken down at the beginning of a unit time, then at the end of the same unit time it may be fixed (i.e. become operational) with probability r_i, or it may remain broken down with probability $\bar{r}_i=1-r_i$.
c) A server may break down at any time, i.e. when it is busy serving, idle, or blocked (first server only).
d) When a server breaks down the customer in service is assumed to be fully served.
e) The maximum number of customers that can be accommodated in the second node is m. The second queue can accommodate at most m-1 customers, and the server can accommodate a customer in front of it whether it is operational or broken down.
f) The first server operates under the blocking-after-service mechanism.

Under the above assumptions, the queueing network can be described by a Markov chain. The state of the system is described at the end of each unit time by the vector (s_1,s_2,n), where n is the number in the second queue and s_i is the state of the ith server, i=1,2. We have n=0,1,...,m, and s_i=0,1,2,3. The first server, at any instant, may be busy serving (s_1=0), broken down (s_1=1), blocked but not broken down (s_1=2), and, blocked and broken down (s_1=3). Similarly, the second server, at any instant, may be busy serving (s_2=0), broken down and having a customer in front of it (s_2=1), operational but without having a customer in front of it (s_2=2), and broken down without having a customer in front of it (s_2=3)

Now, let us group the states of the system so that they all have the same number of customers in the second node. The states within each group are arranged lexicographically. Groups are arranged in increasing order of the number of customers in the second node. For instance, for m=5 we obtain the following arrangement of states.

Customers in node 2	States
0	020, 030, 120, 130
1	000, 010, 100, 110
2	001, 011, 101, 111
3	002, 012, 102, 112
4	003, 013, 103, 113
5	004, 014, 104, 114
6	204, 214, 304, 314

Now let us construct the one-step transition probability matrix $\mathbf{P}$. The non-zero elements of $\mathbf{P}$ have the following familiar block tri-diagonal structure.

$$\mathbf{P} = \begin{pmatrix} \mathbf{A'} & \mathbf{B'} & & & & & \\ \mathbf{C} & \mathbf{A} & \mathbf{B} & & & & \\ & \mathbf{C} & \mathbf{A} & \mathbf{B} & & & \\ & & \mathbf{C} & \mathbf{A} & \mathbf{B} & & \\ & & & \mathbf{C} & \mathbf{A} & \mathbf{B} & \\ & & & & \mathbf{C} & \mathbf{A} & \mathbf{B} \\ & & & & & \mathbf{C''} & \mathbf{A''} \end{pmatrix},$$

where

$$\mathbf{A'} = \begin{pmatrix} 0 & 0 & 0 & 0 \\ 0 & 0 & 0 & 0 \\ r_1 \bar{p}_2 & r_1 p_2 & \bar{r}_1 \bar{p}_2 & \bar{r}_1 p_2 \\ r_1 r_2 & r_1 \bar{r}_2 & \bar{r}_1 r_2 & \bar{r}_1 \bar{r}_2 \end{pmatrix}, \mathbf{B'} = \begin{pmatrix} \bar{p}_1 \bar{p}_2 & \bar{p}_1 p_2 & p_1 \bar{p}_2 & p_1 p_2 \\ \bar{p}_1 r_2 & \bar{p}_1 \bar{r}_2 & p_1 r_2 & p_1 \bar{r}_2 \\ 0 & 0 & 0 & 0 \\ 0 & 0 & 0 & 0 \end{pmatrix}$$

$$\mathbf{C} = \begin{pmatrix} 0 & 0 & 0 & 0 \\ 0 & 0 & 0 & 0 \\ r_1 \bar{p}_2 & r_1 p_2 & \bar{r}_1 \bar{p}_2 & \bar{r}_1 p_2 \\ 0 & 0 & 0 & 0 \end{pmatrix}, \mathbf{A} = \begin{pmatrix} \bar{p}_1 \bar{p}_2 & \bar{p}_1 p_2 & p_1 \bar{p}_2 & p_1 p_2 \\ 0 & 0 & 0 & 0 \\ 0 & 0 & 0 & 0 \\ r_1 r_2 & r_1 \bar{r}_2 & \bar{r}_1 r_2 & \bar{r}_1 \bar{r}_2 \end{pmatrix}$$

$$\mathbf{B} = \begin{pmatrix} 0 & 0 & 0 & 0 \\ \bar{p}_1 r_2 & \bar{p}_1 \bar{r}_2 & p_1 r_2 & p_1 \bar{r}_2 \\ 0 & 0 & 0 & 0 \\ 0 & 0 & 0 & 0 \end{pmatrix}, \mathbf{C}'' = \begin{pmatrix} \bar{p}_1 \bar{p}_2 & \bar{p}_1 p_2 & p_1 \bar{p}_2 & p_1 p_2 \\ 0 & 0 & 0 & 0 \\ r_1 \bar{p}_2 & r_1 p_2 & \bar{r}_1 \bar{p}_2 & \bar{r}_1 p_2 \\ 0 & 0 & 0 & 0 \end{pmatrix}$$

and

$$\mathbf{A}'' = \begin{pmatrix} 0 & 0 & 0 & 0 \\ \bar{p}_1 r_2 & \bar{p}_1 \bar{r}_2 & p_1 r_2 & p_1 \bar{r}_2 \\ 0 & 0 & 0 & 0 \\ r_1 r_2 & r_1 \bar{r}_2 & \bar{r}_1 r_2 & \bar{r}_1 \bar{r}_2 \end{pmatrix}.$$

The stationary probability vector can now be obtained using a numerical procedure (see Chapter 2).

3.4. Phase-type services

In this section, we consider a two-node open queueing network with blocking assuming that the first queue is saturated. The service time at the first and at the second node is assumed to have a phase-type distribution. We present two methods for the analysis of this system, namely, a matrix-recursive method and a matrix-geometric method where matrix **R** can be obtained using a closed-form expression.

A matrix-recursive procedure

Let us consider the two-node queueing network shown in figure 3.1, assuming that the first node is saturated and that the second node has a finite capacity of size m. Blocking-after-service is assumed. That is, the server of the first node gets blocked when a departing customer finds the second node full. The server becomes unblocked when a departure occurs from the second node. For presentation purposes, we first consider the case where the service time distribution at the first and second node is C_2 with parameters λ_1, λ_2, a_1 and μ_1, μ_2, a_2, respectively. Let (s_1,s_2,n) be the state of the system, where s_1=1,2,3, is the state of the server at the first node, s_2=0,1,2, is the state of the server at the second node, and n=0,1,...,m, is the number of customers in the second node. The first server may be, at any time, busy serving in phase 1 (s_1=1), or busy serving in phase 2 (s_1=2), or blocked (s_1=3). The second server may be, at any time, idle (s_2=0), or busy serving in phase 1 (s_2=1), or busy serving in phase 2 (s_2=2). Let $p(s_1,s_2,n)$ be the steady-state probability that the system is in state (s_1,s_2,n). Below, we consider the simple, but not trivial, case where the capacity m of the second node is equal to 3. The steady-state balance equations are

$$\lambda_1 p(1,0,0) = \mu_1(1-a_2)p(1,1,1)+\mu_2 p(1,2,1) \qquad (3.4.1a)$$

$$\lambda_2 p(2,0,0) = \mu_1(1-a_2)p(2,1,1)+\mu_2 p(2,2,1)+\lambda_1 a_1 p(1,0,0) \qquad (3.4.1b)$$

$$(\lambda_1+\mu_1)p(1,1,1) = \lambda_1(1-a_1)p(1,0,0)+\lambda_2 p(2,0,0) +\mu_1(1-a_2)p(1,1,2)+\mu_2 p(1,2,2) \qquad (3.4.2a)$$

$$(\lambda_1+\mu_2)p(1,2,1) = \mu_1 a_2 p(1,1,1) \qquad (3.4.2b)$$

$$(\lambda_2+\mu_1)p(2,1,1) = \lambda_1 a_1 p(1,1,1)+\mu_1(1-a_2)p(2,1,2) +\mu_2 p(2,2,2) \qquad (3.4.2c)$$

$$(\lambda_2+\mu_2)p(2,2,1) = \lambda_1 a_1 p(1,2,1)+\mu_1 a_2 p(2,1,1) \qquad (3.4.2d)$$

$$(\lambda_1+\mu_1)p(1,1,2) = \lambda_1(1-a_1)p(1,1,1)+\lambda_2 p(2,1,1) +\mu_1(1-a_2)p(1,1,3)+\mu_2 p(1,2,3) \qquad (3.4.3a)$$

$$(\lambda_1+\mu_2)p(1,2,2) = \lambda_2 p(2,2,1)+\lambda_1(1-a_1)p(1,2,1) +\mu_1 a_2 p(1,1,2) \qquad (3.4.3b)$$

$$(\lambda_2+\mu_1)p(2,1,2) = \lambda_1 a_1 p(1,1,2)+\mu_1(1-a_2)p(2,1,3)+\mu_2 p(2,2,3) \qquad (3.4.3c)$$

$$(\lambda_2+\mu_2)p(2,2,2) = \lambda_1 a_1 p(1,2,2)+\mu_1 a_2 p(2,1,2) \qquad (3.4.3d)$$

$$(\lambda_1+\mu_1)p(1,1,3) = \lambda_1(1-a_1)p(1,1,2)+\lambda_2 p(2,1,2) +\mu_1(1-a_2)p(3,1,3)+\mu_2 p(3,2,3) \qquad (3.4.4a)$$

$$(\lambda_1+\mu_2)p(1,2,3) = \lambda_2 p(2,2,2)+\lambda_1(1-a_1)p(1,2,2) +\mu_1 a_2 p(1,1,3) \quad (3.4.4b)$$

$$(\lambda_2+\mu_1)p(2,1,3) = \lambda_1 a_1 p(1,1,3) \quad (3.4.4c)$$

$$(\lambda_2+\mu_2)p(2,2,3) = \lambda_1 a_1 p(1,2,3)+\mu_1 a_2 p(2,1,3) \quad (3.4.4d)$$

$$\mu_1 p(3,1,3) = \lambda_1(1-a_1)p(1,1,3)+\lambda_2 p(2,1,3) \quad (3.4.5a)$$

$$\mu_2 p(3,2,3) = \lambda_2 p(2,2,3)+\lambda_1(1-a_1)p(1,2,3)+\mu_1 a_2 p(3,1,3). \quad (3.4.5b)$$

Before we proceed to describe the matrix-recursive procedure, let us define **p**(n), n=0,1,2,3, as the sub-vector of the stationary vector of probabilities that there are n customers in the second node. Let us also define **p**(4) as the sub-vector of the stationary vector of probabilities that there are 3 customers in the second node and the first server is blocked.

Consider now equations (3.4.1a), (3.4.1b), (3.4.2b) and (3.4.2d). These equations can be put in the matrix form

$$(p(1,1,1),p(1,2,1),p(2,1,1),p(2,2,1))\begin{pmatrix} \mu_1(1-a_2) & \mu_1 a_2 & 0 & 0 \\ \mu_2 & -\lambda_1-\mu_2 & 0 & \lambda_1 a_1 \\ 0 & 0 & \mu_1(1-a_2) & \mu_1 a_2 \\ 0 & 0 & \mu_2 & -\lambda_2-\mu_2 \end{pmatrix}$$

$$= (p(1,0,0),p(2,0,0))\begin{pmatrix} \lambda_1 & 0 & -\lambda_1 a_1 & 0 \\ 0 & 0 & \lambda_2 & 0 \end{pmatrix},$$

or $\mathbf{p}(1)\mathbf{C_0}=\mathbf{p}(0)\mathbf{A_0}$, where $\mathbf{C_0}$ and $\mathbf{A_0}$ are the matrix coefficients of **p**(1) and **p**(0) respectively in the above matrix equation. From this we have $\mathbf{p}(1)=\mathbf{p}(0)\mathbf{A_0}\mathbf{C}_\mathbf{0}^{-1}$ or

$$\mathbf{p}(1) = \mathbf{p}(0)\mathbf{Z}(1), \quad (3.4.6)$$

where $\mathbf{Z}(1)=\mathbf{A_0}\mathbf{C}_\mathbf{0}^{-1}$.

Now let us put equations (3.4.2a), (3.4.2c), (3.4.3b) and (3.4.3d) in matrix form. We have

$$(p(1,1,2),p(1,2,2),p(2,1,2),p(2,2,2))\begin{pmatrix} \mu_1(1-a_2) & \mu_1 a_2 & 0 & 0 \\ \mu_2 & -\lambda_1-\mu_2 & 0 & \lambda_1 a_1 \\ 0 & 0 & \mu_1(1-a_2) & \mu_1 a_2 \\ 0 & 0 & \mu_2 & -\lambda_2-\mu_2 \end{pmatrix}$$

$$= (p(1,1,1),p(1,2,1),p(2,1,1),p(2,2,1))\begin{pmatrix} \lambda_1+\mu_1 & 0 & -\lambda_1 a_1 & 0 \\ 0 & -\lambda_1(1-a_1) & 0 & 0 \\ 0 & 0 & \lambda_2+\mu_1 & 0 \\ 0 & -\lambda_2 & 0 & 0 \end{pmatrix}$$

$$+ (p(1,0,0),p(2,0,0))\begin{pmatrix} -\lambda_1(1-a_1) & 0 & 0 & 0 \\ -\lambda_2 & 0 & 0 & 0 \end{pmatrix},$$

or $\mathbf{p}(2)\mathbf{C_0}=\mathbf{p}(1)\mathbf{A_1}+\mathbf{p}(0)\mathbf{B_0}$, where $\mathbf{A_1}$ and $\mathbf{B_0}$ are the matrix coefficients of $\mathbf{p}(1)$ and $\mathbf{p}(0)$ respectively in the above matrix equation. Solving for $\mathbf{p}(2)$ we obtain

$$\mathbf{p}(2) = [\mathbf{p}(1)\mathbf{A_1}+\mathbf{p}(0)\mathbf{B_0}]\mathbf{C_0^{-1}}$$

$$= [\mathbf{p}(0)\mathbf{Z}(1)\mathbf{A_1}+\mathbf{p}(0)\mathbf{B_0}]\mathbf{C_0^{-1}}$$

$$= \mathbf{p}(0)[\mathbf{Z}(1)\mathbf{A_1}+\mathbf{B_0}]\mathbf{C_0^{-1}}$$

or

$$\mathbf{p}(2) = \mathbf{p}(0)\ \mathbf{Z}(2), \tag{3.4.7}$$

where $\mathbf{Z}(2)=[\mathbf{Z}(1)\mathbf{A_1}+\mathbf{B_0}]\mathbf{C_0^{-1}}$.

Likewise from equations (3.4.3a), (3.4.3c), (3.4.4b), (3.4.4d) we have

$$(p(1,1,3),p(1,2,3),p(2,1,3),p(2,2,3))\begin{pmatrix} \mu_1(1-a_2) & \mu_1 a_2 & 0 & 0 \\ \mu_2 & -\lambda_1-\mu_2 & 0 & \lambda_1 a_1 \\ 0 & 0 & \mu_1(1-a_2) & \mu_1 a_2 \\ 0 & 0 & \mu_2 & -\lambda_2-\mu_2 \end{pmatrix}$$

$$= (p(1,1,2),p(1,2,2),p(2,1,2),p(2,2,2))\begin{pmatrix} \lambda_1+\mu_1 & 0 & -\lambda_1 a_1 & 0 \\ 0 & -\lambda_1(1-a_1) & 0 & 0 \\ 0 & 0 & \lambda_2+\mu_1 & 0 \\ 0 & -\lambda_2 & 0 & 0 \end{pmatrix}$$

$$+ (p(1,1,1),p(1,2,1)),p(2,1,1),p(2,2,1))\begin{pmatrix} -\lambda_1(1-a_1) & 0 & 0 & 0 \\ 0 & 0 & 0 & 0 \\ -\lambda_2 & 0 & 0 & 0 \\ 0 & 0 & 0 & 0 \end{pmatrix},$$

or $\mathbf{p}(3)\mathbf{C_0}=\mathbf{p}(2)\mathbf{A_1}+\mathbf{p}(1)\mathbf{B_1}$, where $\mathbf{B_1}$ is the matrix coefficient of $\mathbf{p}(1)$ in the above matrix equation. Solving for $\mathbf{p}(3)$ we have

$$\begin{aligned} \mathbf{p}(3) &= [\mathbf{p}(2)\mathbf{A_1}+\mathbf{p}(1)\mathbf{B_1}]\mathbf{C_0^{-1}} \\ &= [\mathbf{p}(0)\mathbf{Z}(2)\mathbf{A_1}+\mathbf{p}(0)\mathbf{Z}(1)\mathbf{B_1}]\mathbf{C_0^{-1}} \\ &= \mathbf{p}(0)[\mathbf{Z}(2)\mathbf{A_1}+\mathbf{Z}(1)\mathbf{B_1}]\ \mathbf{C_0^{-1}} \end{aligned}$$

or

$$\mathbf{p}(3) = \mathbf{p}(0)\ \mathbf{Z}(3),$$

where $\mathbf{Z}(3)=[\mathbf{Z}(2)\mathbf{A_1}+\mathbf{Z}(1)\mathbf{B_1}]\mathbf{C_0^{-1}}$.

Finally, using equations (3.4.5a) and (3.4.5b) we obtain

$$(p(3,1,3),p(3,2,3))\begin{pmatrix} \mu_1 & -\mu_1 a_2 \\ 0 & \mu_2 \end{pmatrix}$$

$$= (p(1,1,3),p(1,2,3),p(2,1,3),p(2,2,3))\begin{pmatrix} \lambda_1(1-a_1) & 0 \\ 0 & \lambda_1(1-a_1) \\ \lambda_2 & 0 \\ 0 & \lambda_2 \end{pmatrix},$$

or $\mathbf{p}(4)\mathbf{A_3}=\mathbf{p}(3)\mathbf{B_3}$, where $\mathbf{A_3}$ and $\mathbf{B_3}$ are the matrix coefficient of $\mathbf{p}(4)$ and $\mathbf{p}(3)$ respectively in the above matrix equation. From this we have

$$\begin{aligned} \mathbf{p}(4) &= \mathbf{p}(3)\mathbf{B}_3\mathbf{A}_3^{-1} \\ &= \mathbf{p}(0)\mathbf{Z}(3)\mathbf{B}_3\mathbf{A}_3^{-1} \\ &= \mathbf{p}(0)\mathbf{Z}(4), \end{aligned}$$

where $\mathbf{Z}(4)=\mathbf{Z}(3)\mathbf{B}_3\mathbf{A}_3^{-1}$.

We now have a recursive scheme for obtaining $\mathbf{p}(n)$ as a function of $\mathbf{p}(0)$. This recursive scheme requires the knowledge of $\mathbf{Z}(n)$, which is also obtained recursively. There remains only to determine $\mathbf{p}(0)$. This can be obtained from equations (3.4.4a) and (3.4.4c) which were not used so far. These two equations can be expressed in the matrix form

$$(p(1,1,3),p(1,2,3),p(2,1,3),p(2,2,3)) \begin{pmatrix} \lambda_1+\mu_1 & -\lambda_1 a_1 \\ 0 & 0 \\ 0 & \lambda_2+\mu_1 \\ 0 & 0 \end{pmatrix}$$

$$= (p(1,1,2),p(1,2,2),p(2,1,2),p(2,2,2)) \begin{pmatrix} \lambda_1(1-a_1) & 0 \\ 0 & 0 \\ \lambda_2 & 0 \\ 0 & 0 \end{pmatrix}$$

$$+ (p(3,1,3),p(3,2,3)) \begin{pmatrix} \mu_1(1-a_2) & 0 \\ \mu_2 & 0 \end{pmatrix},$$

or $\mathbf{p}(0)\mathbf{Z}(3)\mathbf{A}_2=\mathbf{p}(0)\mathbf{Z}(2)\mathbf{B}_2+\mathbf{p}(0)\mathbf{Z}(4)\mathbf{C}_1$, where $\mathbf{A}_2$, $\mathbf{B}_2$, $\mathbf{C}_1$ are the matrix coefficients of $\mathbf{p}(3)$, $\mathbf{p}(2)$ and $\mathbf{p}(4)$ respectively in the above matrix equation. Letting p(1,0,0)=1, we can solve for p(2,0,0). At the end, we will have to normalize all the probabilities.

The recursive procedure can now be summarized as follows:

$$\mathbf{p}(n) = \mathbf{p}(0)\mathbf{Z}(n), \quad n=1,2,3,4,$$

where

$$\mathbf{Z}(1) = \mathbf{A}_0\mathbf{C}_0^{-1}$$

$$\mathbf{Z}(2) = [\mathbf{Z}(1)\mathbf{A}_1+\mathbf{B}_0]\mathbf{C}_0^{-1}$$

$$\mathbf{Z}(3) = [\mathbf{Z}(2)\mathbf{A}_1+\mathbf{Z}(1)\mathbf{B}_1]\mathbf{C}_0^{-1}$$

$$\mathbf{Z}(4) = \mathbf{B}_3\mathbf{A}_3^{-1}$$

and $\mathbf{p}(0)$ can be found from the equation

$$\mathbf{p}(0)\mathbf{Z}(3)\mathbf{A}_2 = \mathbf{p}(0)\mathbf{Z}(2)\mathbf{B}_2 + \mathbf{p}(0)\mathbf{Z}(4)\mathbf{C}_1.$$

The above matrix recursive scheme can be interpreted in a different way by examining the rate matrix. In particular, the rate matrix $\mathbf{Q}$ of the system under study is given in figure 3.3, where $b_1=1-a_1$ and $b_2=1-a_2$.

The horizontal and vertical lines in figure 3.3 have been drawn in order to identify the sub-matrices that correspond to sub-vectors $\mathbf{p}(0)$, $\mathbf{p}(1)$, $\mathbf{p}(2)$, $\mathbf{p}(3)$, and $\mathbf{p}(4)$. In order to obtain the stationary probability vector we have to solve the system $(\mathbf{p}(0),\mathbf{p}(1),\mathbf{p}(2),\mathbf{p}(3),\mathbf{p}(4))\mathbf{Q}=\mathbf{0}$, or

$$(\mathbf{p}(0),\mathbf{p}(1),\mathbf{p}(2),\mathbf{p}(3),\mathbf{p}(4))\begin{pmatrix} \mathbf{D}_0 & \mathbf{E}_0 & & & \\ \mathbf{F}_0 & \mathbf{D}_1 & \mathbf{E}_1 & & \\ & \mathbf{F}_1 & \mathbf{D}_1 & \mathbf{E}_1 & \\ & & \mathbf{F}_1 & \mathbf{D}_1 & \mathbf{E}_2 \\ & & & \mathbf{F}_2 & \mathbf{D}_2 \end{pmatrix} = \mathbf{0}\ ,$$

where $\mathbf{Q}$ has been written in terms of its sub-matrices. From the above matrix equation we have

$$\mathbf{p}(0)\mathbf{D}_0 + \mathbf{p}(1)\mathbf{F}_0 = \mathbf{0} \tag{3.4.8a}$$

$$\mathbf{p}(0)\mathbf{E}_0 + \mathbf{p}(1)\mathbf{D}_1 + \mathbf{p}(2)\mathbf{F}_1 = \mathbf{0} \tag{3.4.8b}$$

$$\mathbf{p}(1)\mathbf{E}_1 + \mathbf{p}(2)\mathbf{D}_1 + \mathbf{p}(3)\mathbf{F}_1 = \mathbf{0} \tag{3.4.8c}$$

$$\mathbf{p}(2)\mathbf{E}_1 + \mathbf{p}(3)\mathbf{D}_1 + \mathbf{p}(4)\mathbf{F}_2 = \mathbf{0} \tag{3.4.8d}$$

$$\mathbf{p}(3)\mathbf{E}_2 + \mathbf{p}(4)\mathbf{D}_2 = \mathbf{0}. \tag{3.4.8e}$$

Now, it is relatively simple to solve for $\mathbf{p}(n)$, $n=0,1,2,3,4$. Using equation (3.4.8a) we can express $\mathbf{p}(1)$ in terms of $\mathbf{p}(0)$, i.e. $\mathbf{p}(1)= -\mathbf{p}(0)\mathbf{D}_0\mathbf{F}_0^{-1}$. From (3.4.8b) we obtain $\mathbf{p}(2)$ in terms of $\mathbf{p}(0)$, i. e.

$$\mathbf{p}(2) = -[\mathbf{p}(0)\mathbf{E}_0+\mathbf{p}(1)\mathbf{D}_1]\mathbf{F}_1^{-1} = \mathbf{p}(0)[\mathbf{D}_0\mathbf{F}_0^{-1}\mathbf{D}_1-\mathbf{E}_0]\mathbf{F}_1^{-1}.$$

Likewise, from (3.4.8c) we can calculate $\mathbf{p}(3)$ in terms of $\mathbf{p}(0)$, and from (3.4.8d) we can obtain $\mathbf{p}(4)$ in terms of $\mathbf{p}(0)$. Finally, we can calculate $\mathbf{p}(0)$ from (3.4.8e) after we substitute for $\mathbf{p}(3)$ and $\mathbf{p}(4)$.

	100	200	111	121	211	221	112	122	212	222	113	123	213	223	313	323
100	$-\lambda_1$	$\lambda_1 a_1$	$\lambda_1 b_1$													
200		$-\lambda_2$	λ_2													
111	$\mu_1 b_2$		$-\lambda_1-\mu_1$	$\mu_1 a_2$	$\lambda_1 a_1$	$\lambda_1 b_1$										
121	μ_2			$-\lambda_1-\mu_2$		$\lambda_1 a_1$	$\lambda_1 b_1$									
211		$\mu_1 b_2$			$-\lambda_2-\mu_1$	$\mu_1 a_2$	λ_2									
221		μ_2				$-\lambda_2-\mu_2$		λ_2								
112			$\mu_1 b_2$				$-\lambda_1-\mu_1$	$\mu_1 a_2$	$\lambda_1 a_1$		$\lambda_1 b_1$					
122			μ_2					$-\lambda_1-\mu_1$		$\lambda_1 a_1$		$\lambda_1 b_1$				
212					$\mu_1 b_2$				$-\lambda_2-\mu_1$	$\mu_1 a_2$	λ_2					
222					μ_2					$-\lambda_2-\mu_2$		λ_2				
113							$\mu_1 b_2$				$-\lambda_1-\mu_1$	$\mu_1 a_2$	$\lambda_1 a_1$		$\lambda_1 b_1$	
123							μ_2					$-\lambda_1-\mu_1$		$\lambda_1 a_1$		$\lambda_1 b_1$
213									$\mu_1 b_2$				$-\lambda_2-\mu_1$	$\mu_1 a_2$	λ_2	
223									μ_2					$-\lambda_2-\mu_2$		λ_2
313											$\mu_1 b_2$				$-\mu_1$	$\mu_1 a_2$
323											$-\mu_2$					$-\mu_2$

Figure 3.3: The Rate matrix Q of the system under study.

In particular, we can set p(1,0,0)=1 and then solve for p(2,0,0). Finally, all probabilities $p(s_1,s_2,n)$ need to be normalized. Unfortunately, this simple procedure for solving for $\mathbf{p}(n)$ does not work in this case, because matrices $\mathbf{F_0}$, $\mathbf{F_1}$ and $\mathbf{F_2}$ are singular (i.e. they cannot be inverted). Alternatively, we can start from equation (3.4.8e) and then move backwards in order to express $\mathbf{p}(n)$, n=0,1,2,3, in terms of $\mathbf{p}(4)$. In this case, we need to invert the sub-matrices over the diagonal, i. e. $\mathbf{E_0}$, $\mathbf{E_1}$ and $\mathbf{E_2}$, which are also singular!

The matrix recursive method avoids the problem of matrix singularity by re-arranging the columns of the rate matrix $\mathbf{Q}$. The rate matrix with re-arranged columns is shown in figure 3.4. (We note that re-arranging the columns does not change the rate matrix. This is the same as considering the balance equations (3.4.1) to (3.4.5) in a different order.) It is easy to check how these columns have been re-arranged in figure 3.4, since each column is associated with a particular state given at the top of the rate matrix.

The sub-matrices of $\mathbf{Q}$ given in figure 3.4 are identical to the sub-matrices used in the recursive scheme described above. Rate matrix $\mathbf{Q}$ can now be written as

$$\mathbf{Q} = \begin{pmatrix} \mathbf{A_0} & \mathbf{B_0} & & & \\ \mathbf{C_0} & \mathbf{A_1} & \mathbf{B_1} & & \\ & \mathbf{C_0} & \mathbf{A_1} & \mathbf{B_2} & \\ & & \mathbf{C_0} & \mathbf{A_2} & \mathbf{B_3} \\ & & & \mathbf{C_1} & \mathbf{A_3} \end{pmatrix},$$

We now note that sub-matrix $\mathbf{C}$ has become non-singular. Therefore, we can obtain $\mathbf{p}(n)$, n=0,1,...,4, by solving the system $(\mathbf{p}(0),\mathbf{p}(1),\mathbf{p}(2),\mathbf{p}(3),\mathbf{p}(4))\mathbf{Q}=\mathbf{0}$, where $\mathbf{Q}$ is given in figure 3.4, following the method discussed above in connection with equations (3.4.8). This approach is in fact the matrix recursive procedure presented above.

We now proceed to describe the matrix recursive procedure in the general case. Let us consider the two-node queueing network with blocking assuming that the first node is always saturated. Blocking-after-service is assumed, and the capacity of the second node is m. The service times at the first and second server have a phase-type distribution. Let $(\alpha_1,\mathbf{S})$ and $(\alpha_2,\mathbf{T})$ be the representation of the phase-type service at the first and second server respectively, where $\alpha_1=(1,0,...,0)$ and

α_2=(1,0,...,0). The order of (α_1,S) and (α_2,T), i.e. the number of phases, is K_1 and K_2 respectively. Let S_{ij} and T_{ij} be the (i,j)th element of matrix **S** and **T** respectively. The matrix recursive method is applicable to phase-type distributions for which $S_{i,j}>0$ and $T_{i,j}>0$ for all $j \geq i$, while all the other elements of **S** and **T** are zero. An example of this particular phase-type service distribution with four phases is shown in figure 3.5. We see that a customer from phase i can only go to phase j, j>i, with probability a_{ij}, or it may depart with probability a_{i0}.

As before, the state of this system is given by (s_1,s_2,n), where s_1, s_1=1,2,...,K_1+1, is the state of the server at the first node, s_2, s_2=0,1,...,K_2, is the state of the server at the second node, and n, n=0,1,...,m, is the number of customers in the second node. The first server may at any time be busy serving in phase i (s_1=i), i=1,2,...K_1, or blocked (s_1=K_1+1). The second server may at any time be idle (s_2=0), or busy serving in phase i (s_2=i), i=1,2,...,K_2. Let p(s_1,s_2,n) be the steady-state probability that the system is in state (s_1,s_2,n). Define also **p**(n), n=0,1,...,m, as the sub-vector of the stationary probability vector that there are n customers in the second node, and **p**(m+1) as the sub-vector of the stationary probability vector that there are m customers in the second node and the first server is blocked. Then, following similar arguments, we can show that probabilities p(s_1,s_2,n) can be obtained using the same matrix-recursive scheme.

The recursive scheme can be summarized as follows:

$$\mathbf{p}(n)=\mathbf{p}(0)\mathbf{Z}(n), \quad n=1,2,...,m+1,$$

where

$$\mathbf{Z}(1)=\mathbf{A_0}\mathbf{C}_0^{-1}$$

$$\mathbf{Z}(2)=[\mathbf{Z}(1)\mathbf{A_1}+\mathbf{B_0}]\mathbf{C}_0^{-1}$$

$$\mathbf{Z}(n)=[\mathbf{Z}(n-1)\mathbf{A_1}+\mathbf{Z}(n-2)\mathbf{B_1}]\mathbf{C}_0^{-1}, \quad n=2,3,...,m,$$

$$\mathbf{Z}(m+1)=\mathbf{B_3}\mathbf{A}_3^{-1}$$

	100	121	200	221	111	122	211	222	112	123	212	223	113	213	313	323
100	$-\lambda_1$		$\lambda_1 a_1$		$\lambda_1 b_1$											
200			$-\lambda_2$		λ_2											
111	$\mu_1 b_2$	$\mu_1 a_2$			$-\lambda_1-\mu_1$		$\lambda_1 a_1$		$\lambda_1 b_1$							
121	μ_2	$-\lambda_1-\mu_2$		$\lambda_1 a_1$		$\lambda_1 b_1$										
211			$\mu_1 b_2$	$\mu_1 a_2$			$-\lambda_2-\mu_1$		λ_2							
221			μ_2	$-\lambda_2-\mu_2$		λ_2										
112					$\mu_1 b_2$	$\mu_1 a_2$			$-\lambda_1-\mu_1$		$\lambda_1 a_1$		$\lambda_1 b_1$			
122					μ_2	$-\lambda_1-\mu_2$		$\lambda_1 a_1$		$\lambda_1 b_1$						
212							$\mu_1 b_2$	$\mu_1 a_2$			$-\lambda_2-\mu_1$		λ_2			
222							μ_2	$-\lambda_2-\mu_2$		λ_2						
113									$\mu_1 b_2$	$\mu_1 a_2$			$-\lambda_1-\mu_1$	$\lambda_1 a_1$	$\lambda_1 b_1$	
123									μ_2	$-\lambda_1-\mu_1$		$\lambda_1 a_1$				$\lambda_1 b_1$
213											$\mu_1 b_2$	$\mu_1 a_2$		$-\lambda_2-\mu_1$	λ_2	
223											μ_2	$-\lambda_2-\mu_2$				λ_2
313													$\mu_1 b_2$		$-\mu_1$	$\mu_1 a_2$
323													$-\mu_2$			$-\mu_2$

Figure 3.4: The Rate matrix Q with rearranged columns.

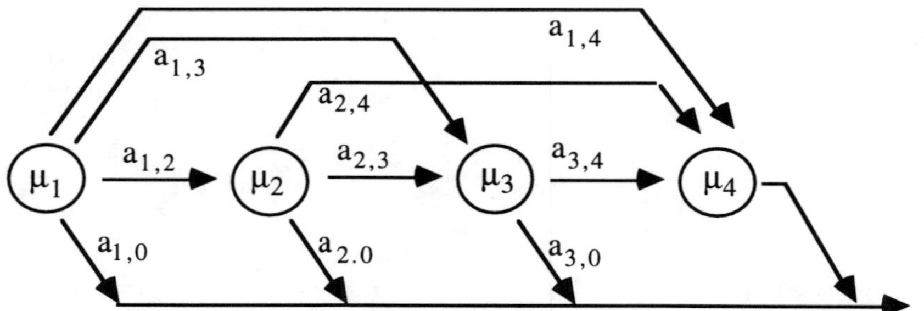

Figure 3.5: The phase-type distribution with four phases.

and $\mathbf{p}(0)$ can be obtained from the equation

$$\mathbf{p}(0)\mathbf{Z}(m)\mathbf{A_2} = \mathbf{p}(0)\mathbf{Z}(m-1)\mathbf{B_2} + \mathbf{p}(0)\mathbf{Z}(m+1)\mathbf{C_1}.$$

Matrices $\mathbf{A_0}$, $\mathbf{B_0}$, $\mathbf{C_0}$,...,$\mathbf{C_1}$ and $\mathbf{A_3}$, are defined in terms of the model parameters. We do so using the Kronecker product of two matrices, $\otimes$. The Kronecker product of matrices $\mathbf{X}$ and $\mathbf{Y}$ is defined as

$$\mathbf{X}\otimes\mathbf{Y} = \begin{pmatrix} x_{11}\mathbf{Y} & x_{12}\mathbf{Y} & \ldots & x_{1M}\mathbf{Y} \\ x_{21}\mathbf{Y} & x_{22}\mathbf{Y} & \ldots & x_{2M}\mathbf{Y} \\ \cdot & \cdot & & \cdot \\ \cdot & \cdot & & \cdot \\ \cdot & \cdot & & \cdot \\ x_{M1}\mathbf{Y} & x_{M2}\mathbf{Y} & \ldots & x_{MM}\mathbf{Y} \end{pmatrix}.$$

Now, matrices $\mathbf{A_0}$, $\mathbf{B_0}$, $\mathbf{C_0}$,...,$\mathbf{C_1}$ and $\mathbf{A_3}$, can be written as

$$\mathbf{A_0} = -\mathbf{S}\otimes\boldsymbol{\alpha}_1$$

$$\mathbf{B_0} = (-\mathbf{S}^o\otimes\boldsymbol{\alpha}_2)\otimes\boldsymbol{\alpha}_1$$

$$\mathbf{C_0} = \mathbf{I}_{K_1 x K_1}\otimes(\mathbf{T}+\mu_1\boldsymbol{\alpha}_2^T\,\boldsymbol{\alpha}_2+\mathbf{T}^o\boldsymbol{\alpha}_2) + \mathbf{S}\otimes(\mathbf{I}-\boldsymbol{\alpha}_2^T\,\boldsymbol{\alpha}_2)$$

$$\mathbf{A_1} = [\mathbf{S}^o\boldsymbol{\alpha}_1\otimes(\mathbf{I}-\boldsymbol{\alpha}_2^T\,\boldsymbol{\alpha}_2)] + (-\mathbf{S}+\mu_1\mathbf{I})\otimes\boldsymbol{\alpha}_2^T\,\boldsymbol{\alpha}_2$$

$$\mathbf{B_1} = (-\mathbf{S}^o\boldsymbol{\alpha}_1)\otimes\boldsymbol{\alpha}_2^T\,\boldsymbol{\alpha}_2$$

$$\mathbf{B_2} = \mathbf{S}^o\otimes\boldsymbol{\alpha}_2^T\,\boldsymbol{\alpha}_2$$

$$A_2 = (-S+\mu_1 I)\otimes\alpha_2^T$$

$$B_3 = S^o\otimes I_{K_2 x K_2}$$

$$C_1 = T^o\alpha_2$$

$$A_3 = -T.$$

The marginal probability distribution $p_2(n)$, n=0,1,...,m, of the number of customers in the second node is

$$p_2(0) = \sum_{i=1}^{K_1} p(i,0,0)$$

$$p_2(n) = \sum_{i=1}^{K_1} \sum_{j=1}^{K_2} p(i,j,n),\ \ n=1,2,...,m-1,$$

$$p_2(m) = \sum_{i=1}^{K_1+1} \sum_{j=1}^{K_2} p(i,j,m).$$

The probability that the first server is blocked is equal to

$$\sum_{j=1}^{K_2} p(K_1+1,j,m).$$

Other performance measures of interest can be easily calculated. For instance, the utilization of node 2 is $[1-p_2(0)](\alpha^T T^{-1} e)$, and the mean number of customers in node 2 is $\sum np_2(n)$.

A matrix-geometric procedure

We will now present an alternative method, based on the matrix-geometric procedure, for analyzing the two-node queueing network studied above. As before, we assume that the first node is saturated and that the second node has a finite capacity of size m. Blocking-before-service is assumed. That is, a customer at the first server does not start service unless there is a space available in the second node. Unlike the case above, the service times at the first and second nodes have an arbitrary phase-type distribution (α_1,S) and (α_2,T), where α_{1i}, $\alpha_{2i} \geq 0$ for all i, and $S_{ij}\geq 0$ and $T_{ij}\geq 0$ for all i and j, $i\neq j$. Let (s_1,s_2,n) be the state of the system, where

s_1=1,2,...,K_1+1, is the state of the server at the first node, s_2=0,1,...,K_2, is the state of the server at the second node, and n=0,1,...,m, is the number of customers in the second node. The first server may be, at any time, busy serving in phase i (s_1=i), i=1,2,...K_1, or blocked (s_1=K_1+1). The second server may be, at any time, idle (s_2=0) or busy serving in phase j (s_2=j), j=1,...,K_2. As above, define **p**(n), n=0,1,...,m, as the sub-vector of the stationary probability vector of that there are n customers in the second node, and **p**(m+1) as the sub-vector of the stationary probability vector that there are m customers in the second node and the first server is blocked. This queueing network can be analyzed using the matrix-geometric procedure described in section 2.5. In particular, the steady-state balance equations can be written as

$$\mathbf{p}(1) = \mathbf{p}(0)\,(\mathbf{S}\otimes\boldsymbol{\alpha}_2)\mathbf{N}^{-1}$$

$$\mathbf{p}(n) = \mathbf{p}(n-1)\mathbf{M}\mathbf{N}^{-1},\ n=1,2,...,m,$$

$$\mathbf{p}(m+1) = -\mathbf{p}(m)\,(\mathbf{S}^{o}\otimes\mathbf{I}_{K_2})\mathbf{T}^{-1}$$

where

$$\mathbf{M} = (\mathbf{I}_{K_1}-\mathbf{e}_{K_1}\alpha_1)\otimes T + \mathbf{S}\otimes I_{K_2}$$

$$\mathbf{N} = \mathbf{S}\otimes(\mathbf{I}_{K_2} - \mathbf{e}_{K_2}\boldsymbol{\alpha}_2) + \mathbf{I}_{K_1}\otimes T$$

and both **M** and **N** are non-singular matrices. We note that in the above solution, **p**(n) is given in terms of **p**(n-1). Thus, **p**(n) can be expressed in terms of **p**(0). We have

$$\mathbf{p}(n) = \mathbf{p}(0)\mathbf{G}\mathbf{R}^{n-1},\ n=1,2,...,m,$$

$$\mathbf{p}(m+1) = \mathbf{p}(0)\mathbf{G}\mathbf{R}^{m-1}\mathbf{U},$$

where **p**(0) is given by

$$\mathbf{p}(0)\mathbf{H} = \mathbf{h}$$

and

$$\mathbf{R} = \mathbf{M}\mathbf{N}^{-1}$$

$$\mathbf{G} = (\mathbf{S}\otimes\boldsymbol{\alpha}_2)\mathbf{N}^{-1}$$

$$\mathbf{U} = - (\mathbf{S}^o \otimes \mathbf{I}_{K_2})\mathbf{T}^{-1}$$

$$\mathbf{H} = \mathbf{W} + \mathbf{G}\mathbf{R}^{m-1}\mathbf{U}\mathbf{e}_{K_2}\mathbf{h}$$

$$\mathbf{W} = \mathbf{I}_{K_1} + [\sum_{k=1}^{m}\mathbf{G}\mathbf{R}^{k-1}] (\mathbf{I}_{K_1} \otimes \mathbf{e}_{K_2})$$

$$\mathbf{h} = \boldsymbol{\alpha}_1(\mathbf{I}_{K_1} - \mathbf{S})^{-1} / \boldsymbol{\alpha}_1(\mathbf{I}_{K_1} - \mathbf{S})^{-1}\mathbf{e}_{K_1}.$$

Notice that the probability that the second node is full, i.e. n=m, is given by $\mathbf{p}(m)\mathbf{e}+\mathbf{p}(m+1)\mathbf{e}$, where $\mathbf{e}$ is a column vector of 1's.

The above matrix-geometric procedure is more general than the matrix-recursive procedure since the service times may have any arbitrary phase-type distribution. Also, computationally, it is slightly more efficient since it requires fewer matrix operations for the calculation of $\mathbf{p}(n)$. The computation of $\mathbf{p}(0)$ is, however, simpler in the case of the matrix-recursive scheme.

3.5. References

The intuitive arguments given for the two limiting cases, the blocking probability, and the condition for stability in section 3.2 are due to Foster and Perros (1980). The two-node queueing network with multiple servers presented at the end of section 3.2 was first analyzed by Makino (1964). The model with constant synchronized service times discussed in section 3.3 was first analyzed by Artamonov (1976). He obtained the closed-form analytic solution by manipulating the steady-state equations algebraically. The matrix-recursive procedure presented in section 3.4 is due to Altiok and Ranjan (1988), and the matrix-geometric procedure is due to Gün and Makowski (1988). Further references related to the topics presented in this Chapter are given below.

Exponential service times

The queueing model discussed in section 3.2 was analyzed by Pellaumail and Boyer (1981). They assumed that the first queue has a finite capacity, and that departing customers from the network may be fed back to the first queue. The rate of external arrivals and the service rate at the first and second node were assumed to be dependent on the state of the system.

The authors devised a recursive procedure for obtaining the queue-length probability distribution. The same queueing network was also analyzed by Konheim and Reiser (1976) (see also Konheim and Reiser (1978)) under blocking-before-service and assuming that a customer departing from the second server may be fed back to the first queue. They used the generating function approach outlined in section 3.2. The authors obtained the necessary additional equations using the fact that the generating function $g_m(z)$ is analytic within the unit circle.

Tsiotras and Badr (1990) and Tsiotras, Badr, and Beltrami (1992) reported on a recursive procedure for obtaining the blocking probability for various special cases. The condition for stability has been derived by several authors (see Asare (1978), Bocharov and Albores (1980), Konheim and Reiser (1976), Latouche and Neuts (1980), Pujolle and Potier (1979), Hunt (1956), and Lavenberg (1978)). We note that Lavenberg (1978) formally demonstrated that when the first queue is saturated, the rate of departures from the queueing network is equal to the supremum of the set of arrivals for which the network is stable. Furthermore, this departure rate is equal to the supremum of the set of departure rates over all finite arrivals.

Latouche and Neuts (1980) investigated the two-node queueing network with multiple servers using the matrix geometric procedure. They considered three different blocking mechanisms based on the following two notions: a) The blocking of a server at the first node may or may not cause the remaining unblocked servers to become blocked; b) Unblocking of a server (or servers) may occur when the number of customers in the second node drops below a predefined number. The authors studied the above queueing system by allowing a feedback loop of departures from the second node to the first node. They observed that the rate matrix of the queueing network under any of the three blocking mechanisms has a block tri-diagonal structure. They also obtained the condition for stability for each of the three different blocking mechanisms. Finally, we note that Makino (1964) also studied the output process of this queueing network.

For further references see Boxma and Konheim (1981), Clarke (1977a), (1977b), (1978a), (1978b), Conway and Keilson (1992), Krishna and Shin (1983), Langaris and Conolly (1984), (1985) and Wong, Giffin, and Disney (1977).

Unreliable servers

Two-node queueing networks with unreliable servers have been studied by Gershwin and Schick (1983), Jafari and Shanthikumar (1985), Muth and Yeralan (1981), and Okamura and Yamashina (1977).

Phase-type services

Shiue and Altiok (1993) analyzed approximately a two-node multi-product production/inventory system with set-ups. Phase-type services and phase-type set-up times were assumed. Neuts (1968) considered a two-node open queueing network assuming general service times at the first server, exponential service times at the second server, and blocking-after-service. He analyzed this model in terms of a semi-Markov process imbedded at instants of service completion at the first server. Most of the results obtained are purely formal. A less general case was considered by Avi-Itzhak and Yadin (1965), Suzuki (1964), and Prabhu (1967) who analyzed its transient behaviour. Pinedo and Wolf (1982) considered two nodes in tandem with and without blocking. They compared the expected waiting time when a customer has independent service times at the two servers with the expected waiting time when a customer has equal service times at the two servers. Rao (1975a) (1975b) obtained the mean effective service time at the first server assuming arbitrary service times, an unlimited supply of customers in the first node and blocking-after-service.

Newell (1979) analyzed two finite capacity nodes in tandem using diffusion approximations. The service times at both nodes and the inter-arrival times were assumed to have the same mean but different variances. Each node was treated as if it were a stochastic continuous fluid. Finally, van Dijk and Lamond (1985) considered a two-node queueing network with blocking where the first node has a finite capacity. They gave a lower and upper bound on the probability that the first node is full using the job-balance property. This property guarantees a product-form solution. The bounds are insensitive to the service distributions for certain service disciplines.

Bibliography

Altiok, T. and Ranjan, R., Analysis of production lines with general service times and finite buffers: A two-node decomposition approach, *Engineering Costs and Production Economics* **17** (1988) 155-165.

Artamonov, G. T., Productivity of a two-instrument discrete processing line in the presence of failures, *Kibernetica* **3** (1976) 126-130 (English tr. *Cybernetics* **12** (1977) 464-468).

Asare, B., *Queue networks with blocking*, Ph.D. Thesis, Trinity College Dublin, Ireland, 1978.

Avi-Itzhak, B. and Yadin, M., A sequence of two servers with no intermediate queue, *Manage. Sci.* **11** (1965) 565-571.

Bocharov, P. P. and Albores, F. K., On two-stage exponential queueing system with internal losses or blocking, *Prob. Control Infor. Theory* **9** (1980) 365-379.

Boxma, O. and Konheim, A., Approximate analysis of exponential queueing system with blocking, *Acta Inf.* **15** (1981) 19-66.

Clarke, A. B., *Markovian queues with servers in tandem*, Tech. Rept. 49, Mathematics Dept., Western Michigan Univ., 1977a.

Clarke, A. B., *A two-server tandem queueing system with storage between servers*, Tech. Rept. 50, Mathematics Dept., Western Michigan Univ., 1977b.

Clarke, A. B., *A multiserver general service time queue with servers in series*, Tech. Rept. 51, Mathematics Dept., Western Michigan Univ., 1978a.

Clarke, A. B., *Waiting times for Markovian queue with servers in series*, Tech. Rept. 52, Mathematics Dept., Western Michigan Univ., 1978b.

Conway, A. E. and Keilson, J., *Delay distribution in a two-stage finite buffer queueing model with blocking and flow control*, Tech. Rept., GTE Labs, 1992.

Foster, F. G. and Perros, H. G., On the blocking process in queue networks, *Eur. J. Oper. Res.* **5** (1980) 276-283.

Gershwin, S. B. and Shick, I. C., Modelling and analysis of three-stage transfer lines with unreliable machines and finite buffers, *Oper. Res.* **31** (1983) 354-380.

Gün, L. and Makowski, A. M., Matrix-geometric solution for finite queues with phase-type distribution, in: Courtois and Latouche, (Eds.), *Performance '87*, (North-Holland, Amsterdam, 1988) 269-282.

Hunt, G. C., Sequential arrays of waiting lines, *Oper. Res.* **4** (1956) 674-683.

Jafari, M. A. and Shanthikumar, J. G., *A model of two-stage flow lines with non-geometric uptimes and downtimes and possible scrapping of workpieces*, Tech. Rept.85-008, Industrial Engineering and Operations Research dept., Syracuse University, 1985.

Konheim, A. G. and Reiser, M., A queueing model with finite waiting room and blocking, *J.ACM* **23** (1976) 328-341.

Konheim, A. G. and Reiser, M., Finite capacity queueing systems with applications in computer modelling, *SIAM J. Comput.* **7** (1978) 210-229.

Krishna, C. and Shin, K., Queueing analysis of a canonical model of real-time multiprocessors, in: Bruell and Dowdy, (Eds.) *Proc. ACM SIGMETRICS Conference on Measurement and Modelling of Computer Systems*, (1983) 175-189.

Langaris, C. and Conolly, B., On the waiting time of a two-stage queueing system with blocking, *J. Appl. Probab.* **21** (1984) 628-638.

Langaris, C. and Conolly, B., Two queues in tandem with multiple facilities and restricted waiting space, *Metron* **43** (1985).

Latouche, G. and Neuts, M. F., Efficient algorithmic solutions to exponential tandem queues with blocking, *SIAM J. of Alg. Disc. Meth.* **1** (1980) 93-106.

Lavenberg, S. S., Stability and maximum departure rate of certain open queueing networks having finite capacity constraints, *RAIRO Informatique/Computer Science* **12** (1978) 353-370.

Makino, T., On the mean passage time concerning some queueing problems of the tandem type, *J. Oper. Res. Soc. Japan* **7** (1964) 17-47.

Muth, E. J. and Yeralan, S., Effect of buffer size on productivity of work stations that are subject to breakdowns, in: *Proc. 20th IEEE Conf. on Decision and Control*, 643-648, 1981.

Neuts, M. F., Two-queues in series with a finite, intermediate waiting room, *J. Appl. Probab.* **5** (1968) 123-142.

Newell, G. F., *Approximate behavior of tandem queues*, Lecture Notes in Economics and Mathematical Systems, **171** (Springer, Berlin, 1979).

Okamura, K. and Yamashina, H., Analysis of the effect of buffer storage capacity in transfer line systems, *AIIE Trans.* **9** (1977) 127-135.

Pellaumail, J. and Boyer, P., *Deux files d'attente à capacité limitée en tandem*, Tech. Rept. 147, CNRS/INRIA - Rennes, 1981.

Pinedo, M. and Wolf, R. W., A comparison between tandem queues with dependent and independent service times, *Oper. Res.* **30** (1982) 464-479.

Prabhu, N. U., Transient behavior of a tandem queue, *Manage. Sci.* **13** (1967) 631-639.

Pujolle, G. and Potier, D., Reseaux de files d'attente a capacité limitée avec des applications aux systèmes informatiques, *RAIRO Informatique/Computer Sci.* **13** (1979) 175-197.

Rao, N. P., On the mean production rate of a two-stage production system of the tandem type, *Int. J. Prod. Res.* **13** (1975a) 207-217.

Rao, N. P., Two-stage production systems with intermediate storage, *AIIE Trans.* **7** (1975b) 414-421.

Shiue, G. A. and Altiok, T., Two-stage, multi-product production/inventory systems, in: Onvural and Akyildiz, (Eds.), *Proc. Second International Workshop on Queueing Networks with Finite Capacity*, (North-Holland, Amsterdam, 1993) 213-223.

Suzuki, T., On a tandem queue with blocking, *J. Oper. Res. Soc. Japan* **6** (1964) 137-157.

Tsiotras, G. D. and Badr, H., A recursive methodology for the derivation of the blocking probabilities of tandem queues with finite capacity, *Computers and Operations Research* **17** (1990) 475-479.

Tsiotras, G. D., Badr, H., and Beltrami, E., Blocking probabilities for a class of two-station tandem queueing models, *IIE Trans.* **23** (1992) 331-345.

van Dijk, N. and Lamond, B. F., Bounds for the call congestion in tandem queues, *Oper. Res.* **36** (1988) 470-477.

Wong, B., Giffin, W. and Disney, R., Two-finite M/M/1 queues in tandem: a matrix solution for the steady state, *Opsearch* **14** (1977) 1-18.

CHAPTER 4

Approximate Analysis of Open Tandem Queueing Networks With Blocking

4.1. Introduction

In this Chapter, we consider open queueing networks consisting of finite capacity queues linked in series. We refer to such an arrangement of queues as a *tandem configuration*. Two-node tandem configurations were discussed in the previous Chapter. Here we consider tandem configurations consisting of more than two nodes. With the exception of some special cases, tandem configurations do not have a closed-form solution. One way of analyzing such queueing networks is to solve numerically for the stationary probability vector of the underlying Markov process (see Chapter 2). Cpu and memory complexity, however, limits this approach to small queueing networks. Alternatively, tandem configurations can be analyzed approximately. In this Chapter, we present several approximation algorithms. These approximation algorithms are based on the notion of decomposition. That is, the tandem configuration is broken up into sub-systems and each sub-system is analyzed separately. In order to analyze a sub-system in isolation, one needs information from other sub-systems. This requirement leads to an iterative scheme whereby the sub-systems are successively analyzed until some criterion of convergence is achieved. The convergence of most of these iterative schemes has been verified empirically, and it has not as yet been proved theoretically.

The approximation algorithms presented below are based on either *single-node* or *two-node* decomposition. In a single-node decomposition, each sub-system consists of a single node; in a two-node decomposition, each sub-system consists of two adjacent nodes. Depending upon its complexity, a sub-system may be analyzed exactly or numerically by solving the underlying Markov process. We note that one can construct a decomposition algorithm where each sub-system consists of more than two adjacent nodes, say three nodes. Such an approximation algorithm will probably be more accurate than a single-node or a two-node decomposition. However, it will be more cpu intensive since it will take longer to analyze each sub-system.

In general, it is quite difficult to carry out a theoretical analysis on the error introduced by such approximation algorithms. In view of this, the accuracy of an approximation algorithm is typically checked by comparing the approximate results against exact numerical results or simulation data. Exact numerical results can be obtained by solving numerically the underlying Markov process of the queueing network under investigation. This method is feasible when the queueing network is not too large. For large queueing networks, simulation appears to be the only feasible method for comparing the approximate results. Comparisons against simulation data can give one a good idea of the accuracy of an approximation algorithm, provided that the simulation data have been obtained using sound statistical methods.

The material in this Chapter is organized as follows. In section 4.2, we analyze tandem configurations where all service times are exponentially distributed. We show how such queueing networks can be analyzed using single-node and two-node decomposition. In section 4.3, we present a single-node and a two-node decomposition algorithm for the analysis of tandem configurations with phase-type service times. In section 4.4, we examine the case of constant service times, and in section 4.5 we discuss the case of synchronized servers subject to breakdowns. Finally, section 4.6 contains relevant references.

4.2. Exponential service times

Let us consider a tandem configuration of M nodes as shown in figure 4.1. For presentation purposes the nodes are numbered from 1 to M starting from the leftmost node. All nodes are assumed to have a finite capacity except the first one which may have an infinite capacity. Let μ_i and m_i be the service rate and maximum capacity of node i, i=1,2,...,M, respectively. All service times are assumed exponentially distributed. Customers in each node are served in a FIFO manner. Customers arrive at

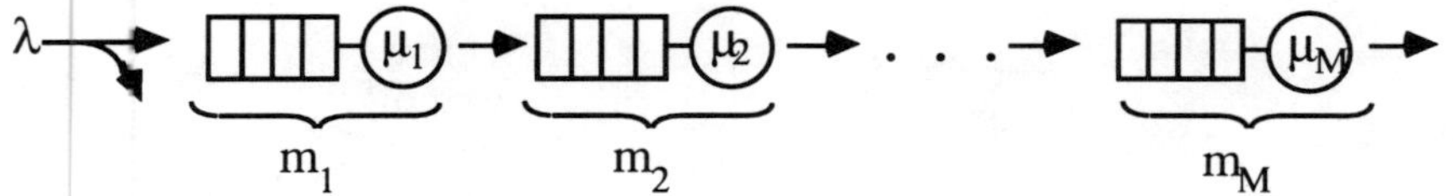

Figure 4.1: An open tandem queueing network with blocking.

the first node in a Poisson fashion at the rate λ. When the first node has an infinite capacity, all arriving customers enter the node. In the case where the first node has a finite capacity, a customer enters the node if it arrives at a time when the node is not full. Otherwise, the customer gets lost.

Let us now consider node i. The nodes prior to node i, i.e. nodes 1 to i-1, are commonly referred to as the *upstream nodes*, and the nodes after node i, i.e. nodes i+1 to M, are referred to as the *downstream nodes*.

We now proceed to analyze this queueing network by first using single-node decomposition and then two-node decomposition. We present two single-node decomposition algorithms and a two-node decomposition algorithm.

Single-node decomposition

We analyze the tandem configuration described above by decomposing it into individual nodes. That is, each sub-system is simply a node. In order to analyze each node in isolation we have to modify its original service process and also characterize its arrival process (see figure 4.2). Also, as it will be seen later on, in the case of blocking-after-service, we have to augment its buffer capacity by one.

Due to the finite capacity of its immediate downstream node i+1, server i will get blocked from time to time. The server remains blocked until a service completion occurs at node i+1. It may also get blocked for longer, if the customer who just completed its service at node i+1 gets blocked by node i+2. It is also possible that when node i gets blocked, nodes i+1 to j, j≥i+1, are already blocked. In this case, when a customer completes its service at node j+1, and given that it does not get blocked, nodes i to j become unblocked and each blocked customer moves up to the next node. This blocking effect due to the downstream nodes has to be incorporated into the service mechanism of each node when studied in isolation. The

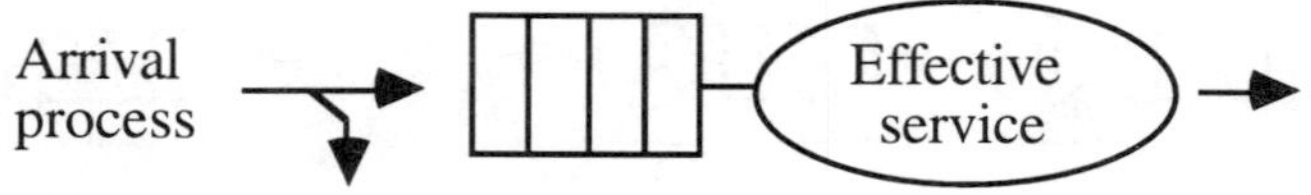

Figure 4.2: Node i studied in isolation.

total time a customer spends in front of server i, i.e. the original service time plus the additional blocking delay, is referred to as the *effective* service time. Finally, in order to analyze node i in isolation we also have to characterize its arrival process. This arrival process is in effect the departure process from its immediate upstream node. Due to blocking, this process depends on the behaviour of the downstream nodes.

Single-node decomposition algorithms are centered round the problem of determining the parameters of the effective service time and of the arrival process to each node studied in isolation. Below, we present two approximation algorithms. We shall refer to them as algorithm 1 and algorithm 2. In both algorithms it is assumed that the arrival process to a node studied in isolation is Poisson. This assumption works well in the case where the service times are exponentially distributed. One of the main difference between these two algorithms is in the way the effective service time is constructed. In the first algorithm the effective service time is assumed to be exponentially distributed, whereas in the second algorithm it is characterized by a phase-type distribution.

a. Algorithm 1: Exponential effective service times

Let us consider the tandem queueing network described above assuming blocking-after-service. That is, after a customer completes its service at the ith node, it proceeds to the (i+1)st node if there is space available. If the (i+1)st node is full at that time, the customer is forced to wait in front of the ith node until a departure occurs from the (i+1)st node. During this time, the ith server remains blocked and it cannot give service to any other customers that might be waiting in the node. The first node is assumed to be always saturated (i.e. never empty). As it was mentioned before, this is a common assumption when studying production systems and it implies that there is an unlimited supply of raw material at the first node. However, it is not an appropriate assumption when studying computer communication systems.

The approximation algorithm is in the form of an iterative scheme, and it yields the system's throughput. Let $p_i(n_i)$ be the marginal probability that there are n_i customers in the ith node. The arrival and service processes for each node studied in isolation are characterized as follows.

We study each node i, i=2,3,...,M, assuming that the arrival process is Poisson, part of which gets lost due to the finite capacity of the node. Let λ_i be the rate of this overall arrival process to node i. The effective service time is also assumed to be exponentially distributed. Let $\bar{\mu}_i$ be the effective service rate of the ith node, i=1,2,...,M. We can now analyze node i as an M/M/1 queue with finite capacity. The capacity of node i is augmented by one in order to accommodate the blocked customer. This is necessary due to the nature of the blocking-after-service mechanism. This is because the (i-1)st server acts as an additional space for the ith node during the time period that it is blocked by this node. Therefore, the blocked customer is in effect part of the ith node, although it waits in front of the (i-1)st server. In view of this, node i can be analyzed in isolation as an $M/M/1/m_i+1$ queue. Let us now see how we can obtain λ_i and $\bar{\mu}_i$.

We recall that the first node is never empty. Therefore, the rate at which customers depart from this node, assuming that none of them get blocked, is μ_1. Therefore, for the second node we have $\lambda_2=\mu_1$. This implies that we allow the first node to continuously serve customers. Those customers that are served during the time that server 1 is supposed to be blocked are simply treated as lost customers from the point of view of the second node. In general, the percentage of time node i-1 is busy (serving or being blocked) is $1-p_{i-1}(0)$. Therefore, for the ith node, i=2,3,...,M, we set

$$\lambda_i = \mu_{i-1}(1-p_{i-1}(0)) . \tag{4.2.1}$$

(For i=2, we have $\lambda_2=\mu_1$ since $p_1(0)=0$). Now, let T be the throughput of the queueing network. Then, we have that $\bar{\mu}_i[1-p_i(0)]=T$, or

$$\bar{\mu}_i = \frac{T}{1-p_i(0)} . \tag{4.2.2}$$

The last node is never blocked. Therefore, we have $\bar{\mu}_M=\mu_M$, and $\mu_M[1-p_M(0)]=T$. Since each node i, i=2,3,...,M, is analyzed in isolation as an $M/M/1/m_i+1$ queue, we have that

$$1-p_i(0) = \frac{\rho_i(1-\rho_i^{m_i+1})}{1-\rho_i^{m_i+2}}, \qquad (4.2.3)$$

where $\rho_i=\lambda_i/\bar{\mu}_i$. Using (4.2.1) and (4.2.2) we obtain

$$\rho_i = \frac{\mu_{i-1}[1-p_{i-1}(0)]}{T/[1-p_i(0)]}. \qquad (4.2.4)$$

Using (4.2.4) into (4.2.3), we can solve for $p_i(0)$ as a fixed-point problem if $p_{i-1}(0)$ and T are known.

The algorithm can now be summarized as follows.

Step 1: Let $T^{(0)}$ be an initial value of the system's throughput.

Step 2: Since $p_1(0)=0$, solve equations (4.2.3) and (4.2.4) as a fixed-point problem in order to obtain $p_2(0)$. Using $p_2(0)$ and $T^{(0)}$ solve for $p_3(0)$ in the same way. Repeat this approach until $p_M(0)$ is calculated.

Step 3: Using $p_M(0)$ compute a new value for T, call it $T^{(1)}$. If $|T^{(0)}-T^{(1)}|<\varepsilon$, stop. Otherwise, set $T^{(0)}=T^{(1)}$ and go to step 1.

Validation tests showed that the approximation algorithm has a good accuracy with an average relative error in the system's throughput T of less than 5%.

b. *Algorithm 2: Phase-type effective service times*

We now proceed to describe an alternative single-node approximation algorithm where the effective service time at each node is represented by a phase-type distribution. The first node is not saturated, and it may have a finite or an infinite capacity. Blocking-after-service is assumed. The queueing network is decomposed into individual nodes with modified arrival and service processes and augmented buffer capacity. Each node is analyzed in isolation starting from node M and working backwards.

As before, we study each node i, i=2,3,...,M, in isolation assuming that the overall arrival process to the node is Poisson distributed, part of which gets lost due to the finite capacity of the node. For node i, let λ_i be the rate of this overall arrival process. Also let $\bar{\lambda}_i$ be the effective arrival rate, that is the rate of customers who actually enter the node. The service mechanism of each server i that is liable to getting blocked is enhanced by appropriately adding exponential phases. The resulting service time distribution is of phase-type and it represents the regular service a customer receives plus any possible delays it might undergo due to blocking. For the same reason given in algorithm 1, the capacity of each node (except node 1) is augmented by one in order to accommodate the blocked customer. Therefore, given that we know λ_i and the parameters of the effective service time we can analyze node i as an $M/PH/1/m_i+1$ queue. In the original implementation of this algorithm the matrix-geometric procedure was used (see Chapter 2).

The main result that can be obtained using this approximation algorithm is the queue-length probability distribution of each node. We first present this algorithm by assuming that the first node has an infinite capacity. Subsequently, we show how it can adjusted for the case where the first node has a finite capacity.

Let us now consider the case where the first node has an infinite capacity. We assume that the system is stable (that is, the first node is stable). Since the arrival rate of customers at the first node is λ, the throughput of the queueing network is also λ. In view of this, when we analyze node i in isolation, i=2,3,...,M, we set its effective arrival rate $\bar{\lambda}_i$ equal to λ. The overall rate of arrivals λ_i is then equal to $\lambda_i=\lambda/[1-p_i(m_i+1)]$.

Now, we proceed to describe how the service mechanism at each node is modified in order to take into account the delay due to blocking. We first note that the Mth server cannot get blocked. Therefore, it can be analyzed in isolation as an $M/M/1/m_M+1$ queue with an arrival rate λ_M. This quantity can be obtained by solving the fixed-point problem $\lambda_M =\bar{\lambda}_M/[1-p_M(m_M+1)]$, where $p_M(m_M+1)=(1-\rho_M)\rho_M^{m_M+1}/(1-\rho_M^{m_M+2})$ and $\rho_M=\lambda_M/\mu_M$.

Let us now consider the (M-1)st node. This node can get blocked by the Mth node. Therefore, it is analyzed in isolation as an $M/C_2/1/m_{M-1}+1$ queue. The two-phased Coxian distribution shown in figure 4.3 reflects the effective service time a customer receives at the (M-1)st server. In particular, a customer starting its service receives first an exponentially distributed service with mean $1/\mu_{M-1}$. This is equivalent to the customer receiving its service at the (M-1)st server. Upon completion of this service the customer receives a second exponentially distributed service with mean $1/\mu_M$ with probability $a_{M-1,M}$. This reflects the event that a customer upon completion of its (M-1)st service finds the Mth node full and, is forced to

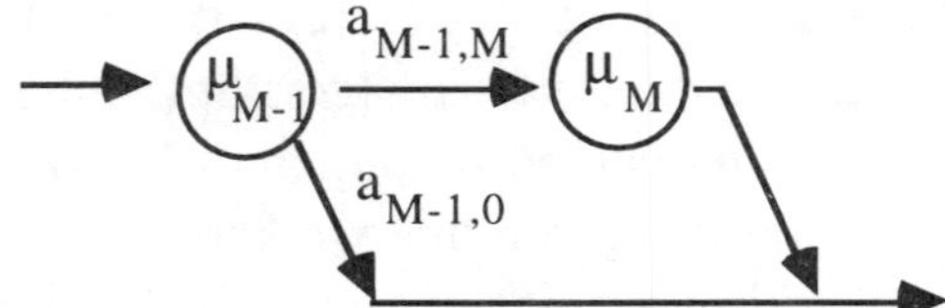

Figure 4.3: The effective service time of the (M-1)st node.

wait until a departure occurs from the Mth server. Obviously, $a_{M-1,M} + a_{M-1,0} = 1$. The quantity $a_{M-1,M}$ is calculated as follows. Let π_{M-1} be the steady-state conditional probability that upon service completion at the (M-1)st node, the Mth node is full (i.e. it contains m_M customers). Then, $a_{M-1,M}=\pi_{M-1}$. The probability π_{M-1} is approximated by the probability that an arrival to the Mth node, when this node is considered in isolation, finds m_M customers and is forced to occupy the (m_M+1)st position. By applying Little's relation to the (m_M+1)st position of the Mth node, considered in isolation, we obtain $(\lambda\pi_{M-1})(1/\mu_M)=p_M(m_M+1)$. The term $\lambda\pi_{M-1}$ is the rate at which an entering customer is forced to occupy this position, $1/\mu_M$ is the mean time a customer spends in this position, and $p_M(m_M+1)$ is the mean number of customers in this position. From the above expression we have

$$\pi_{M-1} = \mu_M p_M(m_M+1)/\lambda, \tag{4.2.5}$$

where $p_M(m_M+1)$ has already been calculated from the analysis of the Mth node.

We analyze the (M-1)st node in isolation as an $M/C_2/1/m_{M-1}+1$ queue using the matrix-geometric procedure. In order to analyze this queue, however, we need to know in advance the overall input rate λ_{M-1}. This quantity is computed numerically using the fixed-point expression $\lambda_{M-1}=\lambda/[1-p_{M-1}(m_{M-1}+1)]$. That is, assuming a value for λ_{M-1} we solve to obtain the queue-length distribution of the (M-1)st node. From this, we calculate a new value for λ_{M-1} and we iterate in this fashion until the successive values of λ_{M-1} converge.

Now, let us consider the (M-2)nd node. This can be modelled as an $M/PH_3/1/m_{M-2}+1$ queue with a service mechanism as shown in figure 4.4. We note that if the first phase is removed, we will obtain the effective

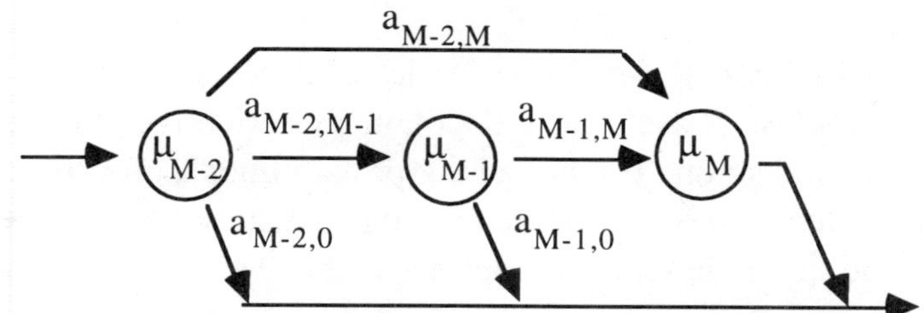

Figure 4.4: The effective service time of node M-2.

service time of the (M-1)st node. A customer starting its service at this phase-type server receives first an exponential service with mean $1/\mu_{M-2}$. This is equivalent to the customer receiving its service at the (M-2)nd server. Upon completion of this service, the customer may depart with probability $a_{M-2,0}$. This is equivalent to the customer completing its service time at the (M-2)nd server at a time the (M-1)st queue is not full. However, the customer will get blocked if the (M-1)st node happens to be full at that instant. The customer will remain blocked, thus blocking the (M-2)nd server, for a period of time which depends on the state of the (M-1)st server at the time that the blocking occurred. The (M-1)st server, at that instant, may be either busy serving or blocked by the Mth node. If it is blocked, the customer will remain blocked for a period of time equal to the residual service time of the Mth server. Obviously, this period of time is exponentially distributed with a mean time of $1/\mu_M$. This is reflected in the phase-type server by allowing the blocking customer to join the last phase. Now, if the (M-1)st server is busy serving, then the customer will remain blocked until a departure occurs from the (M-1)st server. This is reflected in the phase-type server by allowing the customer to join the second phase. In this case, the period of time during which the customer will remain blocked is equal to the residual service time at the (M-1)st server, which has a mean of $1/\mu_{M-1}$, plus a possible delay the (M-1)st server might suffer due to the Mth queue being full at the moment of service completion at the (M-1)st server.

Proceeding in this fashion, the effective service time of each node i can be characterized by a phase-type distribution involving M-i+1 phases. These phases are numbered from i to M, and phase k, k=i, i+1,...,M, has a service rate equal to μ_k. The phases are arranged so that for each phase k, $a_{k0}>0$ and $a_{kj}>0$ for j=k+1,k+2...,M. If we remove the first phase, i.e. phase i, we will obtain the effective service of the (i+1)st node. If we also remove the (i+1)st phase, we will obtain the effective service of the (i+2)nd node, and so on. The branching probabilities a_{ij} for the effective service at the ith node are calculated approximately using information related to the queue-length distribution of the (i+1)st node, which has already been analyzed approximately. As before, let π_i be the steady-state conditional probability that upon service completion at the ith node, the (i+1)st node is full (i.e. it contains m_{i+1} customers). Also, let $w_{i+1}(j)$ be the steady-state probability that the (i+1)st server is in the jth phase of its Coxian effective service time, given that the (i+1)st node is full. Then, the branching probabilities a_{ij} can be approximated by $a_{ij}=\pi_i w_{i+1}(j)$, where j=i+1,i+2...,M, and i≥1. The quantities π_i and $w_{i+1}(j)$ are obtained using the queue-length probability distribution of the (i+1)st node.

Each node i is analyzed as an $M/PH_{M-i+1}/1/m_i+1$ queue using the matrix-geometric procedure. The effective service at the ith node is given by the phase-type distribution $(\boldsymbol{\alpha}_i,\mathbf{T}_i)$, where $\boldsymbol{\alpha}_i$ is a row vector with M-i+1 elements all equal to zero except the first element that is set to 1, and

$$\mathbf{T}_i = \begin{pmatrix} -\mu_i & \mu_i a_{i,i+1} & \mu_i a_{i,i+2} & \cdots & \mu_i a_{i,M-1} & \mu_i a_{i,M} \\ & -\mu_{i+1} & \mu_{i+1} a_{i+1,i+2} & \cdots & \mu_{i+1} a_{i+1,M-1} & \mu_{i+1} a_{i+1,M} \\ & & -\mu_{i+2} & \cdots & \mu_{i+2} a_{i+2,M-1} & \mu_{i+2} a_{i+2,M} \\ & & & & \cdot & \cdot \\ & & & & -\mu_{M-1} & \mu_{M-1} a_{M-1,M} \\ & & & & & -\mu_M \end{pmatrix}.$$

It should be mentioned that in order to analyze each node i, i≥2, one needs to know in advance the total input rate λ_i rather than the effective arrival rate $\bar{\lambda}_i$, where $\bar{\lambda}_i=\lambda$. This quantity can be computed numerically using the fixed-point expression $\lambda_i=\lambda/[1-p_i(m_i+1)]$. That is, assuming a value for λ_i we solve numerically to obtain the queue-length distribution of the ith node. This fixed-point problem is quite similar to the one employed in the previous algorithm. From this we calculate a new value for λ_i and we iterate in this fashion until the successive values of λ_i's converge.

Given that we have analyzed node i, the quantities $w_i(j)$, j=i,i+1,...,M, and i>1, are obtained using the expression

$$w_i(j) = \frac{p_i(0)\, \boldsymbol{\alpha}_i\, \mathbf{R}_i^{m_i}\, \mathbf{e}_j}{p_i(m_i)}, \tag{4.2.6}$$

where $p_i(0)$ is the probability that the ith node is empty, $p_i(m_i)$ is the probability that the ith node contains m_i customers, $\mathbf{R}_i$ is the $\mathbf{R}$ matrix from the matrix-geometric solution, and $\mathbf{e}_j$ is a vector of size M-i+1 with all elements equal to zero except the jth element which is equal to 1. The quantity π_{i-1} is obtained in the same way as above by applying Little's relation in the (m_i+1)st position. We have

$$\lambda\pi_{i-1}\left[\mathbf{w}_i\mathbf{T}_i^{-1}\,\mathbf{e}\right] = p_i(m_i+1)\ , \tag{4.2.7}$$

where $\mathbf{w}_i=(w_i(i),...,w_i(M))$, and $\mathbf{e}$ is an M-i+1 vector with all elements equal to 1. The quantity in the square brackets is the expected remaining service time at the ith node at the instance when the (i-1)st node gets blocked.

The approximation algorithm can be summarized as follows.

Step 1: Analyze the Mth node as an $M/M/1/m_M+1$ queue. Calculate the queue-length probabilities $p_M(n)$. Calculate π_{M-1} using (4.2.5). Set i=1.

Step 2: Analyze the M-i node as an $M/PH_{M-i+1}/1/m_{M-i}+1$ node, i=1,2,…,M-1. Calculate its queue-length probabilities $p_{M-i}(n)$ using the matrix-geometric procedure. Calculate $w_{M-i}(j)$, j=M-i,...,M, using (4.2.6) and $\pi_{M-(i+1)}$ using (4.2.7). Use the results for $w_{M-i}(j)$ and $\pi_{M-(i+1)}$ in order to characterize the phase-type distribution of the effective service time of the (M-(i+1))st node. Add $p_i(m_i)+p_i(m_i+1)$ in order to obtain the actual probability that node M-i is full. Set i=i+1. If i=M-1, go to step 3. Else, repeat step 2.

Step 3: Analyze node 1 as an $M/PH_{M-i+1}/1$ node using the matrix-geometric procedure. Stop.

The above algorithm can be also used when the first node has a finite capacity. In order to use this algorithm, one needs to know in advance the effective input rate $\bar{\lambda}_i$ into each node i, i>1. In the case that the first node has an infinite capacity, this quantity is readily available as it is equal to the arrival rate λ. When the first node has a finite capacity, the effective arrival rate into the first node is not known. In this case, the algorithm can be imbedded into a simple iterative scheme whereby the value of $\bar{\lambda}_1$ is approximated by successive iterations. In particular, let $\bar{\lambda}_1^{(0)}$ be an initial estimate. Using the approximation algorithm, the queue-length distribution of the first node can be calculated, from which a second value $\bar{\lambda}_1^{(1)}$ can be obtained. We stop, if $|\bar{\lambda}_1^{(0)} - \bar{\lambda}_1^{(1)}| < \varepsilon$. Else we set $\bar{\lambda}_1^{(0)} = \bar{\lambda}_1^{(1)}$ and we repeat the above procedure.

Validation tests of this approximation algorithm showed that it has a good accuracy in the case of balanced tandem configurations, i.e. where the service rates are all equal or nearly equal. The observed average relative error in the queue-length distribution and in the mean queue-length of each node was about 5%. This is due to the fact that the coefficient of variation of the departure process from node i is almost equal to 1, and therefore the assumption that the arrival process to a node is Poisson becomes reasonable. Good approximation results were also observed for unbalanced tandem configurations where the first queue has a finite capacity. The observed average relative error in the queue-length distribution and in the mean queue-length of each node was a lttle over 5%. In the case of unbalanced tandem configurations where the first node has an infinite capacity, the approximate results related to the first few nodes tend to have a high relative error. Finally, we note that in the case of large buffers, the relative error for the queue-length distribution gets higher towards the tail of the distribution.

Algorithm 3: Two-node decomposition

In this section, we present a two-node decomposition algorithm for the approximate analysis of tandem queueing networks. The algorithm is applicable to tandem configurations where the first node has a finite capacity, and the arrival rate to the first node is state-dependant, i.e. $\lambda(n_1)$. The service rate at each node is also assumed to be state-dependent, i.e. $\mu_i(n_i)$. Notice that the assumptions regarding the arrival and service rates are more general than those in the previous two algorithms. The approximation algorithm is an iterative procedure. It gives the queue-length distribution at each node, and the joint queue-length distribution for pairs of neighbouring nodes. We present the approximation algorithm assuming blocking-after-service. The tandem configuration is analyzed by decomposing it into sub-systems, each consisting of two adjacent nodes. Sub-system 1 consists of nodes 1 and 2, sub-system 2 consists of nodes 2 and 3, and in general, sub-system i consists of nodes i and i+1, i=1,2,...,M-1. The sub-systems are shown in figure 4.5 for a tandem configuration consisting of 6 nodes. Each node in this figure is represented by a circle. Now, let (n_i,s_i) be the state of node i, where n_i is the number of customers in node i, including the one in service, and s_i is the status of the ith server. s_i=bl if the ith server is blocked, and s_i=nb otherwise. Let $p(n_i,s_i,n_{i+1},s_{i+1})$ be the joint steady-state probability distribution of the number of customers in the ith sub-system, i.e. in nodes i and i+1. This joint distribution can be obtained by solving the underlying Markov process of the sub-system numerically. However, in order to analyze these two nodes we need to know how node i interacts with node i-1, and node i+1 interacts with node i+2. In particular, we need to know the following transition rates:

a. The transition rate from (n_i,s_i,n_{i+1},nb) to $(n_i,s_i,n_{i+1}-1,nb)$. This is the rate at which server i+1 completes a job and the job is not blocked by node i+2.
b. The transition rate from (n_i,s_i,n_{i+1},nb) to (n_i,s_i,n_{i+1},bl). This is the rate at which server i+1 completes a job and the job is blocked by node i+2.
c. The transition rate from (m_i,nb,n_{i+1},s_{i+1}) to $(m_i-1,nb,n_{i+1}+1,s_{i+1})$. This is the rate at which server i completes a service and does not trigger an instantaneous arrival to node i (i.e. node i is not blocking server i-1).
d. The transition rate from (m_i,nb,n_{i+1},s_{i+1}) to $(m_i,nb,n_{i+1}+1,s_{i+1})$. This is the rate at which server i completes a service and triggers an instantaneous arrival to node i (i.e. node i was blocking server i-1).
f. The transition rate from (n_i,s_i,n_{i+1},bl) to $(n_i,s_i,n_{i+1}-1,nb)$. This is the rate at which server i+1 gets unblocked by the (i+2)nd node .
g. The transition rate from (m_i,bl,m_{i+1},s_{i+1}) to (m_i,nb,m_{i+1},nb). This is the rate at which a job leaves node i+1 and triggers an instantaneous arrival to node i (i.e. node i and i-1 were blocked).

These transition rates are approximated using the conditional probabilities $p(n_{i-1},s_{i-1}|n_i,s_i)$ and $p(n_{i+2},s_{i+2}|n_{i+1},s_{i+1})$. $p(n_{i-1},s_{i-1}|n_i,s_i)$ gives the probability that the (i-1)st node is in a particular state, given that

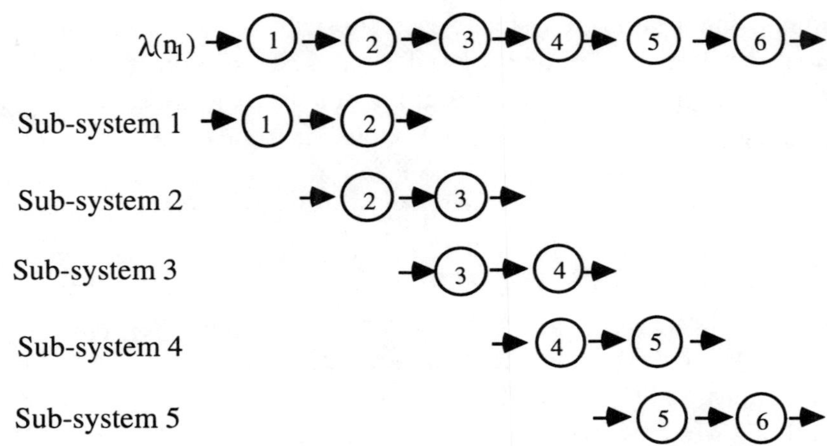

Figure 4.5: Sub-systems in the two-node decomposition algorithm.

we know the state of the ith node; $p(n_{i+2},s_{i+2}|n_{i+1},s_{i+1})$ gives the probability that the (i+2)nd node is in a particular state, given that we know the state of the (i+1)st node. These conditional probabilities are not known, but they are approximated by successive iterations. The steps of the iterative procedure are as follows:

Step 1: Select an initial approximation for all the conditional probabilities $p^{(0)}(n_{i-1},s_{i-1}|n_i,s_i)$ and $p^{(0)}(n_{i+2},s_{i+2}|n_{i+1},s_{i+1})$.

Step 2: (Iteration j). Solve M-1 pairs of neighbouring nodes (i,i+1) for i=1,2,...,M-1. For each pair (i,i+1) use $p^{(j-1)}(n_{i-1},s_{i-1}|n_i,s_i)$ and $p^{(j-1)}(n_{i+2},s_{i+2}|n_{i+1},s_{i+1})$ and solve the sub-system in order to obtain $p^{(j)}(n_i,s_i,n_{i+1},s_{i+1})$. From this, compute probability $p^{(j)}(n_i,s_i|n_{i+1},s_{i+1})$ and $p^{(j)}(n_{i+1},s_{i+1}|n_i,s_i)$ which will be used in the next iteration.

Step 3: If the maximum absolute value between two iterates for each node is less than a given value, then convergence has been attained. Otherwise, set j to j+1 and return to step 2.

Validation tests showed that the approximation algorithm has a better accuracy than algorithm 2. The approximation algorithm, however, has a poor accuracy when there is a big difference in the probability that a node is full between two neighbouring nodes. The fact that the two-node decomposition algorithm has a better accuracy than the single-node decomposition algorithm 2 is not surprising, since each sub-system contains more information than the single-node sub-systems. However, it is more cpu intensive than the single-node decomposition algorithm.

4.3. Phase-type service times

In this section, we discuss two approximation algorithms for analyzing tandem configurations with phase-type service times. We refer to these algorithms as algorithm 4 and 5. In both algorithms, the effective service time at the ith node, i=1,2,…,M-1, is represented by a phase-type distribution which is constructed following similar arguments as in algorithm 2. The main difference between algorithms 4 and 5 is that algorithm 4 is based on single-node decomposition, whereas algorithm 5 is based on two-node decomposition.

Algorithm 4: Single-node decomposition

This algorithm was developed for an open tandem configuration with blocking-after-service, shown in figure 4.6, where each server has a C_2 service time distribution. All nodes, except the first one, are assumed to have a finite capacity. The first node may have a finite or an infinite capacity. Let m_i be the capacity of the ith node, i=1,2,...,M. Also, let $(\mu_{i,1},\mu_{i,2},a_i)$ be the parameters of the C_2 service distribution of the ith node $\mu_{i,j}$ is the service rate at the jth phase of the C_2 distribution, j=1,2, and a_i is the probability that upon service completion at the first phase the customer will move to the second phase of the service. The arrival process to the first node is assumed to be Poisson. It can be easily extended to a phase-type distribution.

The algorithm is based on algorithm 2 presented above. That is, the system is decomposed into individual nodes, and each node is then analyzed in isolation. The capacity of each decomposed node is augmented by one, its service process is modified so that it accounts for the additional blocking delays, and the arrival process is described by a C_2 distribution (rather than a Poisson distribution). The algorithm yields the steady-state queue-length distribution of each node. Let us assume for the moment that the first node has an infinite capacity. The algorithm consists of the following three steps.

Step 1: This is identical to algorithm 2. In this step, the arrival process at each decomposed node i, i=2,3,…,M, is approximated by a Poisson distribution with parameter λ_i such that the effective arrival rate at this node is equal to λ, the system's throughput. The algorithm begins with the analysis of the Mth node and proceeds backwards to node 1. Unlike algorithm 2, the effective service time at each node is collapsed to a C_2 distribution. As an example, let us consider the (M-1)st node. Upon service completion at node M-1, the customer will enter node M without getting blocked with probability $1-\pi_{M-1}$. Otherwise, with probability π_{M-1} it will get blocked. The time it spends being blocked is, of course, the remaining service time at node M at the instant of blocking. When the (M-1)st node gets blocked, the Mth server is busy serving a customer and it is either in phase 1 or phase 2 with probability $w_{M,1}$ and $w_{M,2}$ respectively. If it is in phase 1, the distribution of the remaining service time is the original service time distribution, i.e. a C_2 distribution with parameters $(\mu_{M,1},\mu_{M,2},a_M)$. If it is in phase 2, the remaining service time is distributed exponentially with a mean of $1/\mu_{M,2}$. Hence, the effective service time of the (M-1)st node is described by the phase-type distribution shown in figure 4.7. The quantities π_{M-1} and $w_{M,j}$ are calculated approximately, as in algorithm 2, using information from the Mth queue. That is, $w_{M,j}=p_M(m_M,j)/p_M(m_M)$, where $p_M(m_M,j)$ is the probability that the node is full and the server is in phase j, j=1,2, and $p_M(m_M)$ is the probability that the node is full. The probability π_{M-1} is calculated using Little's relation on the fictitious (m_M+1)st position. We have

$$\lambda\pi_{M-1}\left[w_{M,1}\left(\frac{1}{\mu_{M,1}}+\frac{a_M}{\mu_{M,2}}\right)+w_{M,2}\frac{1}{\mu_{M,2}}\right]=p_M(m_M+1).$$

The quantity in the square brackets in the above expression approximates the residual service time at the Mth server at a time when the (M-1)st server gets blocked. Now, using the method of moments this

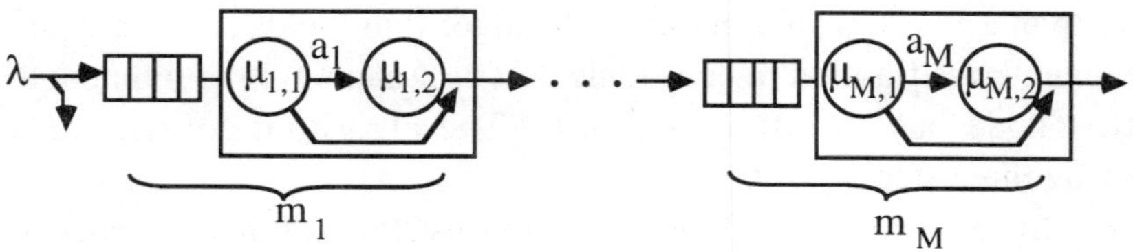

Figure 4.6: A tandem configuration with C_2 service times.

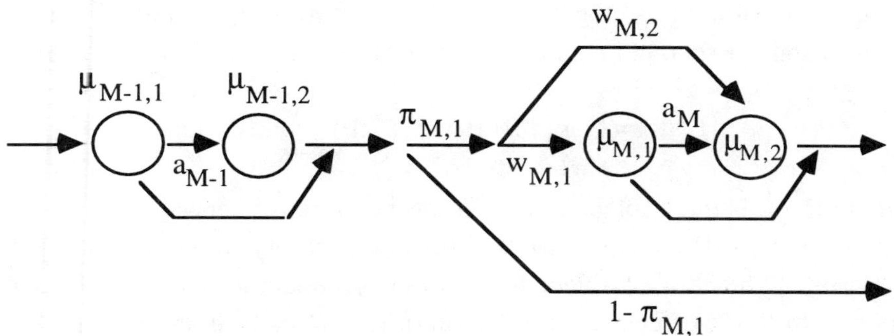

Figure 4.7: The effective service time of the (M-1)st server.

phase-type distribution is approximated by a C_2 distribution. This simplifies the effective service time without compromising the algorithm's accuracy. Node M-1 can now be analyzed as an $M/C_2/1/m_{M-1}+1$ queue using the matrix-geometric procedure. In order to do this, we need to know in advance the overall arrival rate λ_{M-1} rather than the effective arrival rate λ. As in the algorithm 2, λ_{M-1} can be found solving the fixed-point problem $\lambda_{M-1}=\lambda/[1-p_{M-1}(m_{M-1}+1)]$.

Working backwards we can analyze each node i in isolation as an $M/C_2/1/m_i+1$, except node 1 which is analyzed as an $M/C_2/1$ node.

Step 2: This step deals with approximating the arrival process of each decomposed node by a C_2 distribution, while keeping the effective service process the same as modified in step 1. As before, the capacity of each node is augmented by one. The analysis of the network proceeds from node 1 to node M, and each node is analyzed in isolation using an algorithm by Yao and Buzacott (1984).

Let us first consider node 2. In order to characterize its arrival process, one needs to know the pdf of the inter-departure time from node 1. This, in general, is difficult to obtain. Instead, the Laplace transform $D_1^*(s)$ of the inter-departure time from node 1 is obtained assuming that successive effective services are independent of each other. When a customer departs from node 1, either another customer is available in the node ready to begin service, or the node becomes empty. In the first case, the time until the next departure from node 1 is distributed exactly as the effective service time. In the second case, the time until the next departure is the sum of two intervals, the first being the time until the next customer arrives and the second being the customer's service time. Thus, we have

$$D_1^*(s) = (1-p_1^+(0))\, B_1^*(s) + p_1^+(0)\, \frac{\lambda}{s+\lambda}\, B_1^*(s),$$

where $B_1^*(s)$ is the Laplace transform of the effective service time at node 1, and $p_1^+(0)$ is the probability that node 1 is empty immediately after a departure. The above Laplace transform is set equal to $E_2^*(s)$, the Laplace transform of the effective inter-arrival time at node 2. In order to analyze this node, one needs to know the overall arrival process rather than the effective arrival process. The Laplace transform $O_2^*(s)$ of the inter-arrival time of the overall arrival process is given by the expression

$$E_2^*(s) = \frac{(1-f_2)O_2^*(s)}{1-f_2 O_2^*(s)},$$

where f_2 is the probability that an arriving customer finds node 2 full and is lost. This is the probability that an arrival sees m_2+1 customers in the node. It can be obtained by conditioning appropriately on the time average state probabilities. From the above expression, the first three moments of the overall arrival process $m_{o2,1}$, $m_{o2,2}$, $m_{o2,3}$ can be obtained in terms of f_2 and the first three moments of the effective arrival process $m_{e2,1}$, $m_{e2,2}$, $m_{2,3}$. We have

$$m_{o2,1} = (1-f_2)m_{e2,1}$$

$$m_{o2,2} = (1-f_2)(m_{e2,2}-2f_2 m_{e2,1}^2)$$

$$m_{o2,3} = (1-f_2)(m_{e2,3}-6f_2 m_{e2,2} m_{e2,1}+6f_2^2 m_{e2,1}^3).$$

Using these three moments we can construct a C_2 representation of the overall arrival process. Subsequently, node 2 can be analyzed as a $C_2/C_2/1/m_2+1$ queue. We note, however, that f_2 is still undetermined. This can be obtained as in step 1 by solving the fixed-point problem $1/m_{o2,1}=\lambda(1-f_2)$. This guarantees that the throughput is equal to λ.

The algorithm proceeds forward analyzing each node in the same fashion, until the Mth node is analyzed. The expression of the Laplace transform of the inter-departure time of node i, i≥2, is slightly more complicated than the one given above, since the arrival process to node i is a C_2 rather than Poisson.

Step 3: This step is identical to step 1. The objective is to re-calculate the effective service time at each node. The analysis proceeds from node M-1 to node 1. The effective service time of node i is calculated following the same arguments as in step 1. However, due to the fact that the arrival process at node i+1 is C_2, the quantities $w_{i+1,j}$ are calculated using probabilities conditioned on the time immediately before an arrival. These conditional probabilities can be easily obtained by conditioning appropriately on the time average probabilities.

The approximation algorithm can be summarized as follows. Carry out steps 1, 2 and 3, and then the test for convergence. In particular, let

$$p_i^{(2)}(n_i) \text{ and } p_i^{(3)}(n_i),\ n_i=0,1,\ldots,m_i,$$

be the queue-length distribution of the ith node obtained from step 2 and 3 respectively. If

$$\max_{n_i}|p_i^{(2)}(n_i)-p_i^{(3)}(n_i)|<\varepsilon,$$

then stop. Else, repeat steps 2 and 3 and test for convergence again.

The case where the first node has a finite capacity is treated in a similar way as in algorithm 2. Convergence of the effective arrival rate at the first node is tested after step 1 and subsequently after step 4.

In general, the approximation algorithm gives results which are significantly more accurate than those obtained using only step 1, i.e. under the assumption that the arrival process to each decomposed node is Poisson. The average relative error of the approximate queue-length distribution and mean queue-length of each node was observed to be about 5%, whereas the average relative error from step 1 was observed to be about 13%. The algorithm was also used to analyze tandem configurations with exponential service times. For balanced cases, no significant improvement over the results obtained using step 1 was observed. Some improvement, however, was observed for unbalanced cases.

Finally, in this algorithm we approximate various phase-type distributions by two-phase Coxian distributions using the method of moments. This method can be applied when the c^2 of a phase-type distribution is greater than one. In the original implementation of this algorithm, the two moment approximation, described in section 1.5, was used when the c^2 of a phase-type distribution was less than one.

Algorithm 5: Two-node decomposition

The algorithm presented in this section was developed for an open tandem configuration with blocking-after-service, saturated first node, and phase-type service times. The phase-type distribution has the following special structure. A customer from phase i can only go to a phase j if j>i, or it can depart. An example of this particular phase-type service distribution with four phases is shown in figure 3.5.

The algorithm decomposes the tandem configuration into sub-systems consisting of two successive servers and the in-between buffer. In figure 4.8 we show these sub-systems for a tandem configuration consisting of six nodes. We note that sub-system i consists of servers i-1 and i, and the in-between buffer. There is no sub-system 1. In view of the structure of the sub-system, this algorithm should, strictly speaking, be classified as a *two-server* decomposition, rather than a two-node decomposition!

Each sub-system i is analyzed in isolation after the service time at the (i-1)st and ith servers is modified. In order to modify the service time, we need to know if upon service completion at the ith server the customer will get blocked and how long it would remain blocked. Also, we need to know if upon service completion at the (i-1)st server, the server becomes idle and how long it remains idle. Thus, in the ith sub-system, the modified services of the (i-1)st and ith servers represent the effective arrival process to the ith node and its effective service time respectively. The parameters of the effective service time of the (i-1)st server and of the ith server are obtained using information from the (i-1)st and the (i+1)st sub-system respectively. The following iterative scheme is employed. The algorithm starts with the 2nd sub-system and proceeds forward until it analyses the Mth sub-system. It then executes a backward pass starting from the (M-1)st sub-system back to the 2nd sub-system. In the forward pass, the parameters of the first server of each sub-system are updated, while in the backward pass the parameters of the second server of each sub-system are updated. The algorithm iterates backwards and forwards until the throughputs of all the sub-systems are sufficiently close to each other. It yields the queue-length distribution of each node.

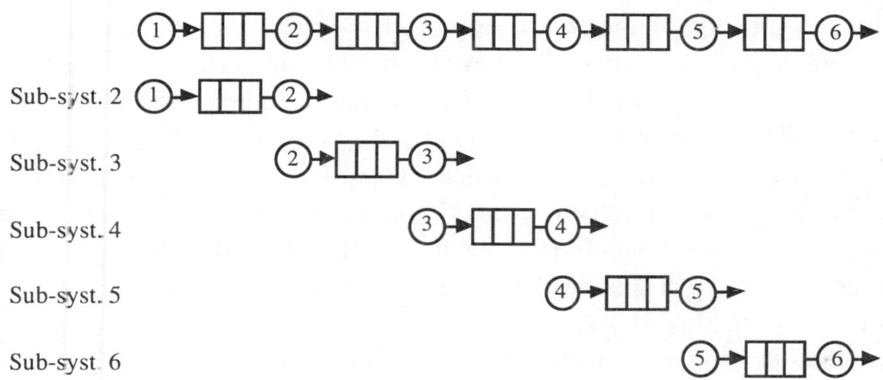

Figure 4.8: The sub-systems in algorithm 4. (Adopted from Altiok and Ranjan (1988))

For simplicity, we shall refer to server i of any sub-system i as the downstream server, and to server i-1 as the upstream server. We now proceed to discuss how the effective service times at the upstream and downstream servers of a sub-system are constructed. For presentation purposes, we shall consider an open tandem configuration where each server has a C_2 service time distribution. The parameters of the service times are shown in figure 4.6. The first node is assumed to be saturated. Let us consider the ith sub-system. The effective service time at the ith server is constructed following the same approach as in algorithm 4. For computational efficiency, the part of the phase-type distribution that represents the blocking delay is approximated by a C_2 distribution using the method of moments. The branching probabilities are calculated using information from the (i+1)st sub-system.

Now, let us consider how the upstream server of each sub-system is modified. In sub-system 2, the upstream server has the same service time distribution as the server of node 1. This is because node 1 is assumed to be saturated and therefore it never becomes idle. Let us now focus on the upstream server of sub-system 3. The effective service time involves the original service at node 2 plus an idle delay that node 2 may experience. In the original tandem configuration, after a customer departs from node 2, the time until the next service completion depends on whether the departing customer leaves node 2 empty or not. If it does not leave node 2 empty, then the elapsed time until the next service completion is equal to the service time of the next customer. However, if it leaves an empty system behind, then node 2 will experience an idle period until a service completion takes place at node 1. The length of the idle period depends on which phase the server in node 1 was in when the customer departed from node 2. Let δ_2 be the probability that a departing customer leaves node 2 empty. We note that a customer may complete its service at a server, but it may not depart from the server due to blocking. δ_2 is the probability that node 2 is empty at the moment when the customer leaves the node, i.e. after its blocking is over, if it was blocked. Define $\beta_2(i)$ to be the probability that the upstream server of sub-system 2 is in its ith phase at the time when the downstream server becomes idle, i=1,2. Then, the modified service time of the upstream server of sub-system 3 has the same structure shown in figure 4.9.

In order to be able to analyze sub-system 3, we need to know $\beta_2(i)$, i=1,2, and δ_2. These quantities are obtained from sub-system 2 as follows. $\beta_2(i)$, i=1,2, is obtained by appropriately conditioning on the state of the system. δ_2 is obtained using Little's relation in the same way as the π_i's were obtained in algorithm 3. We have

$$\bar{\lambda}_2 \, \delta_2 \left[\beta_2(1)\left(\frac{1}{\mu_{1,1}} + \frac{a_1}{\mu_{1,2}}\right) + \frac{\beta_2(2)}{\mu_{1,2}} \right] = p_2(0),$$

where the terms in the square brackets represent the expected idle time server 2 experiences. $p_2(0)$ is the arbitrary time probability that server 2 is idle and it is obtained from the joint distribution of sub-system 2. In order to analyze sub-system 3, we also need to know π_3 and $w_4(i)$ so that we can construct the effective service time. As it was mentioned above, these two quantities are obtained from sub-system 4. In the first forward pass, due to lack of information regarding the sub-systems, all π_i's are assumed to be zero. In the original implementation of this algorithm the sub-system was analyzed using the matrix recursive procedure presented in section 3.4.

Now, let us consider how the effective service process at the upstream server of sub-system 4 is obtained. $\beta_3(i)$ and δ_3, defined as above, can be obtained from sub-system 3 following similar arguments. This is done considering the fact that when the downstream server of sub-system 3 becomes idle, the upstream server of sub-system 3 is in phase i of the phase-type distribution shown in figure 4.9 with probability $\beta_3(i)$. The

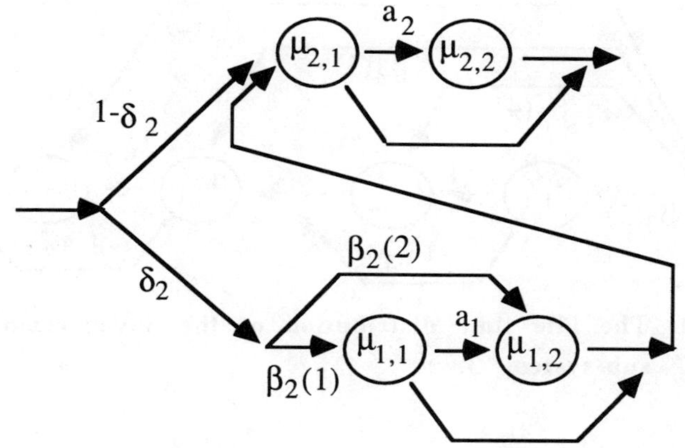

Figure 4.9: The modified service time distribution of the upstream server in subsystem 3

phase-type representation of the idle period that the downstream server in sub-system 3 experiences is given in figure 4.10.

The effective service time at the upstream server of sub-system 4 can now be calculated by combining figure 4.10 with probability δ_3 and the original service time of the third server with probability $1-\delta_3$. For convenience, the phase-type distribution that represents the idle time is approximated by a C_2 distribution using the method of moments. The forward pass continues until the analysis of the (M-1)st sub-system is completed. Then, the backward pass starts with sub-system M. The procedure continues until the throughputs of all the sub-systems are sufficiently close.

In the case where the service times have a C_2 distribution, algorithms 4 and 5, despite their apparent differences, are in fact similar and they only differ in their implementation. Algorithm 4 is more cpu intensive than algorithm 5, since the solution of each node in isolation requires solving a fixed-point problem.

Finally, in this algorithm we approximate various phase-type distributions by two-phase Coxian distributions using the method of moments. This method can be applied when the c^2 of a phase-type distribution is greater than one. In the original implementation of this algorithm, the two moment approximation, described in section 1.5, was used when the c^2 of a phase-type distribution was less than one.

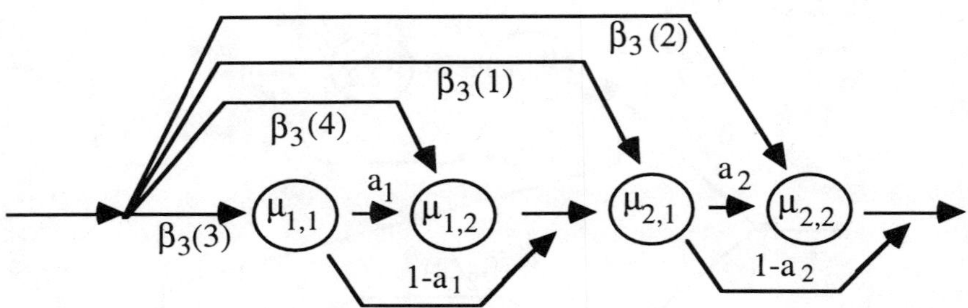

Figure 4.10: The idle time distribution of the downstream server in sub-system 3.

4.4. Constant service times

In this section, we examine tandem configurations with constant service times. Each server provides a constant service time whose length is node-dependent. We distinguish between the case where the first node has an infinite capacity and the case where it has a finite capacity. In the first case, we present exact results for the waiting time of a customer in the system. These results cannot be extended to the case where the first node has a finite capacity. For this case, we only give bounds for the throughput.

First node has an infinite capacity

Let us consider an open tandem configuration consisting of M nodes. Blocking-after-service is assumed, and the service times are constant. That is, server i provides a constant service time of length Y_i, i=1,2,...,M. Customers arrive at the first node, which is assumed to have an infinite capacity, according to an arbitrary arrival process. Each node is served by a single server, and customers are served in a FIFO manner. The following two theorems hold.

Theorem 1: The time spent in the system by any customer is independent of the order of the nodes and of the capacity of nodes 2,3,...,M.

Theorem 2: The total time a customer spends in the whole network is equal to the sum of all the service times plus the total time the customer spends waiting in the various queues of the network. This total waiting time is equal to the time the same customer would spend waiting in a single server queue. The arrival process to this queue is the same as in the queueing network, and the service time is equal to the largest constant service time of the tandem configuration.

These two theorems will be proved below assuming that each node i, except node 1, has a capacity equal to 1, that is, $m_i=1$, $i=2,3,\ldots,M$. The two theorems are immediately applicable to the case where each node i has a capacity $m_i \geq 1$, $i=2,3,\ldots,M$. This is because of the simple observation that each waiting space in a node can be seen as being a server with zero service time. Therefore, a tandem configuration where the first node has an infinite capacity and the remaining nodes have a capacity $m_i \geq 1$, $i=2,3,\ldots,M$, can be seen as a tandem configuration consisting of the same first node linked to $m_2+m_3+\ldots+m_M$ nodes each with capacity 1.

Let k^* be the node with the longest service time. Clearly, a customer will never experience blocking at each of the downstream nodes k^*+1, $k^*+2,\ldots,M$. This is because the time between two successive departures from node k^* is greater than any of the downstream services. Therefore, each time a customer completes service in one of these nodes, the next downstream node is always empty. Hence, no matter what the arrival process at the tandem queueing network is, each customer spends $Y_{k^*+1}+Y_{k^*+2}+\ldots+Y_M$ time in the system after node k^*.

Let us now define the parameters A_n, D_n, S_n and W_n to be, respectively, the arrival time at the queueing network of the nth customer, the departure time from the queueing network of the nth customer, the total time the nth customer spent in the system, and the total time the nth customer spent waiting for service. We note that W_n is equal to the time a customer may spend in the first node queueing for service plus all the blocking delays the customer may experience at servers $1,2,\ldots,M-1$. We have the following relations.

$$S_n = D_n - A_n \qquad (4.4.1)$$

$$W_n = S_n - \sum_{i=1}^{M} Y_i = D_n - A_n - \sum_{i=1}^{M} Y_i. \qquad (4.4.2)$$

We prove theorem 1 by first showing that the following relationship holds for the nth customer, $n \geq 1$.

$$D_n = \begin{cases} A_n + \sum_{i=1}^{M} Y_i \text{ if } W_n = 0 \\ \\ D_{n-1} + Y_{k^*} \text{ if } W_n > 0. \end{cases} \tag{4.4.3}$$

The first part of the above expression is obtained from (4.4.2) by setting $W_n=0$. This expression simply states that if the nth customer does not wait for service, i.e. it does not spend any time queueing for service at node 1 and it does not get blocked at any server, then the total time it spends in the system is equal to the sum of all the services it receives. Therefore, the nth customer departs $Y_1+Y_2+...+Y_M$ units of time after it arrives at the queueing network.

The second part of the expression is proved below and it states that, if the nth customer has to wait for service, then it will depart Y_k* time units after the departure of the (n-1)st customer. Define d_i to be the difference between the total amounts of service already received by the (n-1)st customer and the nth customer at the point of departure of the (n-1)st customer from node i. The total amount of service received by the (n-1)st customer at this point is $Y_1+Y_2+...+Y_i$. We show that the second expression of (4.4.3) is true by considering at which point the nth customer is delayed for the first time. This may happen when the nth customer arrives at node 1 to find the (n-1)st customer in node 1 either waiting for service or receiving service. Alternatively, it may happen when the nth customer upon service completion at server 1,2,...,M-1 gets blocked. That is, at that moment the (n-1)st customer is receiving service at the next downstream server.

Let us consider the first case where the nth customer arrives at node 1 to find the (n-1)st customer also at node 1, either waiting for service or receiving service. The nth customer enters server 1 when the (n-1)st customer departs from server 1 making $d_1=Y_1$. Let us now calculate d_2, the difference between the total amounts of service already received by the (n-1)st customer and the nth customer at the point of departure of the (n-1)st customer from node 2. When the (n-1)st customer departs from the second node, its total amount of service is equal to Y_1+Y_2. The value of d_2 depends on how much service the nth customer has received at server 1 at the moment when the (n-1)st customer departs from node 2. This of course depends on the values of Y_1 and Y_2, or more generally, on the values of d_1 and Y_2. If $Y_2 \geq d_1$, then the nth server will complete its service at server 1 before the (n-1)st customer completes its service at server 2. The nth customer will then get blocked until the (n-1)st customer departs from server 2. Therefore, when the (n-1)st customer leaves server 2, the total amount of service the nth customer has received is equal to Y_1. Thus, if $Y_2 \geq d_1$, then we have that $d_2 = Y_2$.

Now, let us consider the case where $Y_2 \leq d_1$. That is, the (n-1)st customer completes its service before the nth customer completes its service at server 1. The total amount of service the (n-1)st customer has received at the moment it departs from server 2, is $Y_1 + Y_2$. Now, let us see how much service the nth customer receives up to the moment the (n-1)st customer departs from server 2. If the (n-1)st customer does not get blocked by server 3 upon service completion, then the nth customer receives an amount of its service equal to Y_2. Thus, $d_2 = (Y_1 + Y_2) - Y_2 = Y_1$, and we have $Y_2 \leq d_2 = d_1$. Now, if the (n-1)st customer gets blocked upon service completion at server 2, then the nth customer has a chance to complete more of its service up to the maximum possible amount of Y_1. Let x be the amount of service the nth customer completes. Then, $d_2 = (Y_1 + Y_2) - x$, where $Y_2 < x \leq Y_1$. That is, $d_2 < Y_1$ and $d_2 \geq Y_2$ and the following relation holds: $Y_2 \leq d_2 < d_1$. Thus, when $Y_2 \leq d_1$ we have $Y_2 \leq d_2 \leq d_1$. In general, for any server i, i=1,2,...,M, we have

$$\begin{cases} d_i = Y_i & \text{if } Y_i \geq d_{i-1} \\ Y_i \leq d_i \leq d_{i-1} & \text{if } Y_i \leq d_{i-1}. \end{cases} \qquad (4.4.4)$$

Let us now examine the relationship between Y_{k^*} and d_{k^*-1}. We have

$$d_{k^*-1} \leq \max\{Y_1, Y_2, \ldots, Y_{k^*-1}\} \leq Y_{k^*} .$$

Thus, from the first expression of (4.4.4), $Y_{k^*} \geq d_{k^*-1}$ indicates that $d_{k^*} = Y_{k^*}$. That is, when the (n-1)st customer departs from server k^*, the difference in the amount of service received by him and the nth customer is exactly Y_{k^*}. This difference will be maintained when the (n-1)st customer departs from the k^*+1 server. Thus, in general we will have

$$d_{k^*} = d_{k^*+1} = \ldots = d_M = Y_{k^*}.$$

Therefore, at the moment the (n-1)st customer departs from the system, $d_M = Y_{k^*}$. This means that the nth customer has yet to complete Y_{k^*} time units of service. That is, the nth customer will depart Y_{k^*} units of time after the (n-1)st customer departs.

Now, let us consider the case where the nth customer is delayed for the first time because of blocking at the jth server, where $j \leq k^*-1$. This means that the (n-1)st customer is in the (j+1)st server. Therefore, when this customer departs from the (j+1)st server, we will have that $d_{j+1}=Y_{j+1}$. Hence, expression (4.4.4) is again valid for i=j+2,j+3...,M. This completes the proof of validity of (4.4.3).

It should be noted that $d_1=d_2=\ldots=d_M \geq Y_{k^*}$ if $W_n=0$, which includes the case that the nth customer may have arrived some time after the arrival of the (n-1)st customer. In this case, the difference in the amount of service received by the customer n and n-1 will be greater than Y_{k^*}. Then, due to the impossibility of blocking in nodes k^*, $k^*+1,\ldots,M$, we have for $W_n=0$

$$D_n \geq D_{n-1} + Y_{k^*}. \tag{4.4.5}$$

Using (4.4.3) and (4.4.5) we can write

$$A_n + \sum_{i=1}^{M} Y_i \geq D_{n-1} + Y_{k^*}$$

or

$$A_n + \sum_{i \neq k^*}^{M} Y_i \geq D_{n-1}\ . \tag{4.4.6}$$

Replacing the condition $W_n=0$ in (4.4.3) by (4.4.6) we obtain

$$D_n = \begin{cases} A_n+\sum_{i=1}^{M} Y_i & \text{if } A_n+ \sum_{i \neq k^*}^{M} Y_i \geq D_{n-1} \\ D_{n-1}+Y_{k^*} & \text{otherwise.} \end{cases} \tag{4.4.7}$$

This verifies that D_n is determined by D_{n-1}, A_n, Y_{k^*} and the sum $Y_1+Y_2+\ldots+Y_M$. Clearly, D_1 is independent of the order of servers in the system since there are no customers ahead of him. Since A_2, Y_{k^*} and the sum $Y_1+Y_2+\ldots+Y_M$ are independent of the order of servers, we conclude from (4.4.7) that D_2 is also independent of the order of servers. Following this argument, we conclude that D_n is independent of the order of servers for $n \geq 1$. Hence, since D_n and A_n are independent of the order of servers, S_n is also independent of the order because of (4.4.1). This completes the proof of Theorem 1. Notice that addition of more nodes with zero service time has no impact on D_n as can be seen from (4.4.4).

Theorem 2 can be easily proved using theorem 1. For, since we can re-arrange the servers, we let $k^*=1$. That is, no blocking will be experienced in any other server. In this case, node 1 behaves as if it is a single node with a fixed service time Y_{k^*}. Then, the total time the nth customer spends in the queueing network is

$$Wn = W_n(1) + \sum_{i=1}^{M} Y_i \tag{4.4.8}$$

where $W_n(1)$ is the delay experienced by the nth customer in the first node. This proves Theorem 2.

Let

$$\bar{W} = \lim_{n\to\infty} \frac{1}{n} \sum_{i=1}^{n} W_i$$

be the steady-state average time a customer spends in the queueing network, and let

$$W(1) = \lim_{n\to\infty} \frac{1}{n} \sum_{i=1}^{n} W_i(1)$$

be the steady-state average waiting time of a customer in node 1 with a constant service time equal to Y_{k^*}. Then (4.4.8) can be written as

$$\bar{W} = W(1) + \sum_{i=1}^{M} Y_i \ . \tag{4.4.9}$$

In order to obtain $\bar{W}$ we have to compute W(1). In the case where the arrival process is Poisson with rate λ, the first node is an M/D/1 queue and the well-known Pollaczek-Khinchin formula can be used to calculate W(1). That is,

$$W(1) = \frac{\rho Y_{k^*}}{2(1-\rho)},$$

where $\rho=\lambda Y_{k^*}$ is the traffic intensity. In the case where the arrival process is not Poisson but the inter-arrival times are id. random variables with an arbitrary distribution, W(1) can be calculated using the following approximate formula for the average waiting time in a GI/GI/1 queue due to Kramer and Lagenbach-Belz (1976)

$$W(1) = \frac{\rho^2(c_a^2+c_s^2)}{2\lambda(1-\rho)}\, g(c_a^2,c_s^2,\rho)\ , \qquad (4.4.10)$$

where c_a^2 and c_s^2 are the squared coefficient of variation of the inter-arrival times and service times respectively, and

$$g(c_a^2,c_s^2,\rho) = \begin{cases} \exp\left(\dfrac{-2(1-\rho)(1-c_a^2)^2}{3\rho(c_a^2+c_s^2)}\right), & \text{if } c_a^2<1 \\ \\ \exp\left(\dfrac{-(1-\rho)(c_a^2-1)}{c_a^2+4c_s^2}\right), & \text{if } c_a^2\geq 1. \end{cases}$$

Finally, using the fact that the steady-state throughput at each node is equal to λ, the steady-state average number of customers in the tandem system is

$$\lambda W(1) + \lambda\sum_{i=1}^{M} Y_i\ .$$

First node has a finite capacity

The above two theorems were proved under the assumption that the first node has an infinite capacity. In the case where the first node has a finite capacity, the queueing network becomes a loss system. A re-arrangement of the order of nodes or a change in their buffer capacities will have an impact on the number of arrivals entering the system and, therefore, will change the total time a customer spends in the system. This queueing network is difficult to analyze. Below we present a lower and upper bound on the network's throughput.

Let us consider an open tandem configuration consisting of M nodes. Each node has a finite capacity m_i, i=1,2…,M. Blocking-after-service is assumed, and the service times are constant, i.e. server i provides a constant service time of length Y_i, i=1,2,...,M. Customers arrive at the first node, which is assumed to have a finite capacity, in a Poisson fashion at the rate λ. A customer is lost if it arrives to the first node at a time when the node is full. Each node is served by a single server, and customers are served in a FIFO manner. As before, let k^* be the node with the longest service. We will focus on nodes 1 to k^* seeing that after node k a customer is not blocked at any of the subsequent downstream nodes.

Let T be the throughput of the queueing network. Obviously, $0<T\leq\lambda$. We note that 0 and λ can be used as lower and upper bound respectively, though, these two bounds may not be tight. We now introduce the following two bounds.

Lower bound: A lower bound, T_L, is the output rate of an $M/D/1/m_1$ queue with a service time equal to Y_{k*} and an arrival rate equal to λ. That is, T_L is the throughput of the first node if we set the first node's service equal to the maximum service Y_{k*}.

It can be easily shown that this is a lower bound, by simply observing that the maximum possible time that the server at the first node may be blocked is equal to $Y_{k*}-Y_1$. Therefore, the output rate of the first node with a service equal to Y_{k*} is clearly the slowest possible rate at which customers will depart from node 1. Seeing that there will be no blocking in the subsequent downstream nodes, this will also be the slowest possible rate at which customers depart from the queueing network.

Upper bound. An upper bound, T_U, is the minimum of T_0 and $1/Y_{k*}$, where T_0 is the output rate of an $M/D/1/m_1$ queue with a rate of arrivals equal to λ and a constant service time equal to Y_1.

Let us consider the first node in isolation from the remaining network. This node will never get blocked, and therefore its throughput T_0 will be higher than its throughput if the node was part of the network. Because of this, the throughput of the first node treated in isolation from the network may serve as an upper bound for the network's throughput. However, the network's throughput can not be greater than the rate of the slowest server, that is $1/Y_{k*}$. Hence, an upper bound can be $T_U=\min\{T_0,1/Y_{k*}\}$.

These two bounds can be computed by solving an M/D/1 finite capacity queue (see one of the queueing theory textbooks.)

4.5. Synchronized servers

In this section, we examine tandem configurations where all servers are synchronized. The first node is assumed to be saturated, that is, there is always at least one customer waiting to be served. All servers begin and end service at the same time. The duration of a service is constant and it is taken as the unit time. A customer requires a deterministic amount of service at each server which is equal to one unit of time. If we assume that the servers never break down, then there will be no customers waiting in nodes 2 through M. Variability in the service is introduced by allowing a server to break down with some probability. In particular, server i, i=1,2,...,M, breaks down during a unit time with probability p_i. The time to fail is thus distributed geometrically with a mean $1/p_i$. Server i is repaired during a unit time with probability r_i. The time to repair is, therefore, also geometrically distributed with a mean time $1/r_i$. It is assumed that a server may only break down during the time it is busy serving. An idle or blocked server never fails. A customer in service is not lost if during its service the server breaks down. By convention, repairs and failures occur at the beginning of a unit time, and a departure (or an arrival) from (to) a node takes place at the end of a unit time. Blocking-before-service is assumed. That is, server i cannot start a new service if node i+1 is full. This type of queueing network arose from studies of synchronized production assembly systems. It is a very popular model in the area of production systems, and it is often referred to as the *transfer line model*. Below, we discuss an approximation algorithm for the analysis of this type of queueing network. The algorithm yields the throughput of the network and the mean number of customers in each node i, i=2,3,...,M.

Let us consider a tandem configuration of M nodes, and let m_i be the buffer capacity at each node i, i=2,3,...,M. The first node is assumed to be saturated. Each server i is completely characterized by its breakdown and repair probabilities p_i and r_i respectively. The queueing network is analyzed approximately by decomposing it into sub-systems. Each sub-system i, i=2,3,...,M, comprises of servers i-1 and i and the queue in-between. (This type of decomposition is the same as the one used in algorithm 5.) In sub-system i, server i will be hereafter referred to as the downstream server, and server i-1 as the upstream server. The breakdown and repair probabilities of each of these two servers are modified so that sub-system i can be used to approximate the behaviour of node i.

Let $p_u(i)$, $r_u(i)$, and $p_d(i)$, $r_d(i)$ be the breakdown and repair probabilities for the upstream respectively downstream server of the ith sub-system. Let T(i) be the throughput of the ith sub-system. Finally, let $p_0(i)$ and $p_f(i)$ be the probability that there are no customers in the ith sub-system respectively that the ith sub-system is full. An example of the sub-systems and their parameters is given in figure 4.11 for a queueing network of four servers.

The approximation algorithm utilizes the following expressions.

a. *Balanced throughputs*

$$T(2) = T(3) = \ldots = T(M) \tag{4.5.1}$$

This expression simply states that each sub-system has the same throughput. This is due to the assumption that a customer in service is not lost if during its service the server breaks down.

b. *Breakdown-repair probabilities expression*

Let o_i be the percentage of time server i is operational, when the server is considered in isolation. We have $o_i = r_i/(r_i+p_i)$. The throughput of sub-system i can be approximated by the expression

$$T(i) = o_i(1-p_0(i)-p_f(i+1)).$$

Alternatively, using sub-system i we can express T(i) as

$$T(i) = \frac{r_d(i)}{r_d(i)+p_d(i)}(1-p_0(i)).$$

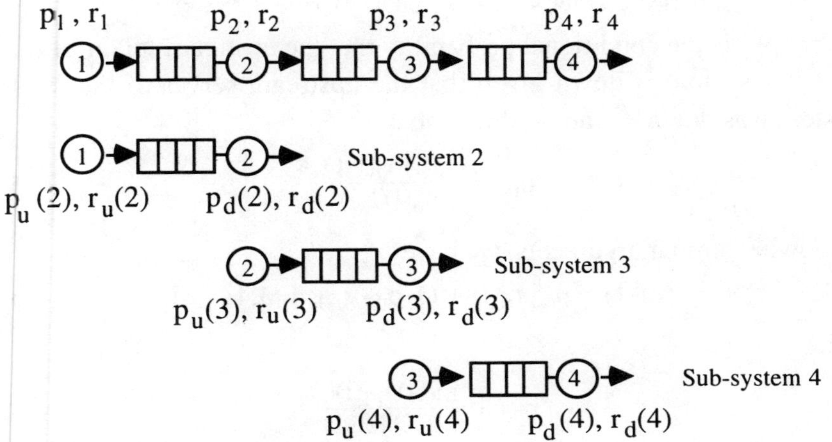

Figure 4.11: An example of the sub-systems and their parameters. (Adopted from Dallery, David, and Xie (1988a))

Finally, from sub-system i+1 we have

$$T(i+1) = \frac{r_u(i+1)}{r_u(i+1)+p_u(i+1)}(1-p_f(i+1)).$$

Combining the above three expressions and using (4.5.1), we obtain

$$\frac{p_d(i)}{r_d(i)} + \frac{p_u(i+1)}{r_u(i+1)} = \frac{1}{T(i)} + \frac{1}{o_i} - 2, \ i=2,3,...,M-1. \qquad (4.5.2)$$

c. *Repair probabilities expression*

A set of expressions for the repair probabilities $r_u(i)$ and $r_d(i)$ can be obtained by examining the meaning of failure and repair of an upstream and a downstream server. An upstream server of a sub-system stops serving customers when it breaks down, or when it is forced to become idle due to the fact that the upstream node is empty. In both cases, the server will appear to the sub-system as being broken down. Therefore, the repair probability $r_u(i)$ is the probability of termination of whichever condition was in effect. That is, it is r_{i-1} if the cause of failure was the failure of the server. Otherwise, it is equal to the probability that the (i-1)st sub-system becomes non-empty. Now, this sub-system became empty because its upstream server broke down. Therefore, the probability that the (i-1)st sub-system becomes non-empty is $r_u(i-1)$. Thus, we have

$$r_u(i) = p_{uu}r_u(i-1) + (1-p_{uu})r_{i-1}, \ i=3,4,...,M, \qquad (4.5.3)$$

where p_{uu} is the conditional probability that the upstream server of the (i-1)st sub-system is down given that the upstream server of the ith sub-system was down. It can be shown that

$$p_{uu} = \frac{r_u(i)p_0(i-1)}{p_u(i)T(i)}. \qquad (4.5.4)$$

Following similar arguments, we have

$$r_d(i-1) = p_{dd}\, r_d(i) + (1-p_{dd})r_{i-1}, \ i=3,4,...,M, \qquad (4.5.5)$$

where

$$p_{dd} = \frac{r_d(i-1)p_f(i)}{p_d(i-1)T(i-1)}. \qquad (4.5.6)$$

Finally, we have the following boundary conditions

$$\begin{cases} p_u(2) = p_1 \\ r_u(2) = r_1 \\ p_d(M) = p_M \\ r_d(M) = r_M \,. \end{cases} \tag{4.5.7}$$

Equations (4.5.1) to (4.5.7) can be slightly re-written so that the queueing network under study can be solved iteratively. Let us first define the quantities

$$I_u(i) = \frac{p_u(i)}{r_u(i)}$$

$$I_d(i) = \frac{p_d(i)}{r_d(i)}$$

for i=2,3,...,M. (4.5.2) can be re-written as

$$I_u(i+1) = \frac{1}{T(i)} + \frac{1}{o_i} - I_d(i)-2, \tag{4.5.8}$$

for i=2,3,...,M-1. Shifting i+1 to i, we have

$$I_u(i) = \frac{1}{T(i-1)} + \frac{1}{o_{i-1}} - I_d(i-1)-2, \tag{4.5.9}$$

for i=3,4,...,M. The conditional probability p_{uu} given by (4.5.4) can be written as

$$p_{uu} = \frac{I_u(i)p_0(i-1)}{T(i-1)}, \tag{4.5.10}$$

where T(i) was substituted by T(i-1). Thus, parameters $r_u(i)$ and $I_u(i)$, and, therefore $p_u(i)$, of the upstream server of the ith sub-system can be obtained from the parameters of the (i-1)st sub-system in terms of equations (4.5.9), (4.5.3) and (4.5.10).

Similarly, parameters $r_d(i)$ and $I_d(i)$, and therefore $p_d(i)$, of the downstream server of the ith sub-system, can be obtained from sub-system i+1 as follows. Substituting T(i) by T(i-1) in (4.5.8) gives

$$I_d(i) = \frac{1}{T(i+1)} + \frac{1}{o_i} - I_u(i+1)-2, \qquad i=2,3,...,M-1. \tag{4.5.11}$$

Changing i-1 to i in (4.5.5) and in (4.5.6) we obtain

$$r_d(i) = p_{dd}\, r_d(i+1) + (1-p_{dd})r_i, \; i=2,3,...,M-1, \tag{4.5.12}$$

$$p_{dd} = \frac{I_d(i)\, p_f(i+1)}{T(i+1)}, \tag{4.5.13}$$

where in (4.5.13) T(i) was substituted by T(i+1). Thus, (4.5.11), (4.4.12) and (4.5.13) can be used to obtain the parameters of the downstream server of the ith sub-system from sub-system i+1. We note that each sub-system was analyzed using an algorithm by Gershwin and Shick (1983).

The algorithm can be summarized as follows:

Step 1: Initialize $p_d(i)=p_i$, $r_d(i)=r_i$, i=2,3,...,M, $p_d(1)=p_1$, $r_d(1)=r_1$.

Step 2: For i=3,4,...,M, calculate $I_u(i)$, $r_u(i)$ and $p_u(i)$, using equations (4.5.9), (4.5.3) and (4.5.10).

Step 3: For M-1,M-2,...,2, calculate $I_d(i)$, $r_d(i)$ and $p_d(i)$ using equations (4.5.11), (4.5.12) and (4.5.13).

Step 4: Stop if convergence has occurred, else go to step 2.

Experimentation with this algorithm has shown that it always converges and it is quite accurate. The average relative error in the estimation of the network's throughput and the mean queue-length of each node was reported to be less than 5%.

4.6. References

Algorithm 1 is due to Hillier and Boling (1967) and it is probably the first approximation algorithm for the analysis of tandem configurations reported in the literature. Algorithm 2 is due to Perros and Altiok (1986), and it is based on an earlier algorithm by Altiok (1982). The two-node decomposition algorithm 3 is due to Brandwajn and Jow (1985) and (1988). In this algorithm, each sub-system was analyzed using a numerical method due to Brandwajn (1979). A speed up technique for algorithm 3 was proposed by Brandwajn and Sahai (1989). Algorithm 4 is due to Jun and Perros (1990). Algorithm 5 is due to Altiok and Ranjan (1988) and it is in principle identical to the algorithm by Gün and Makowski (1989). We note that in the latter algorithm, the service time at each node may follow any arbitrary phase-type distribution. The two theorems and their proofs given in section 4.4 for the case where the first node has an infinite capacity are due to Avi-Itzhak (1965). See also Avi-Itzhak and Levy (1993). The upper and lower bounds given in section 4.4

for the case where the first node has a finite capacity are due to Altiok and Kao (1989). The algorithm presented in section 4.5 for the case of synchronized servers is a modified version of an algorithm by Gershwin (1987a). The modified algorithm is due to Dallery, David and Xie (1988). The approximation algorithms presented in this Chapter can be seen as being different approximation techniques for aggregating large Markov chains. A survey of some of the issues in exact and approximate aggregation is given in Schweitzer (1989). Further references related to the topics presented in this Chapter are given below.

Tandem configurations with exponential service times

Caseau and Pujolle (1979) reported on a single-node decomposition algorithm for obtaining the maximum throughput of a tandem configuration, assuming that the first node has an infinite capacity. Repetitive-service blocking was assumed. The algorithm decomposes the queueing network into individual nodes with revised arrival and service process. Each node is then studied in isolation. The arrival and service processes for node i, i=1,2,...,M, are revised as follows. The arrival process is assumed to be Poisson distributed with an effective rate λ or a total input rate $\lambda/[1-p_i(m_i)]$. The service time is assumed to be exponentially distributed with parameter $\mu_i[1-p_{i+1}(m_{i+1})]$. This parameter is obtained as follows. A customer in node i, upon completion of its service, attempts to enter node i+1. The customer will repeat its service if node i+1 is full at that time. The probability that the customer will repeat its service is approximated by $p_{i+1}(m_{i+1})$, i.e. the probability that node i+1 is full. Therefore, the probability that the customer will repeat its service k times is $[1-p_{i+1}(m_{i+1})][p_{i+1}(m_{i+1})]^{k-1}$. Then, the probability density function of the effective service time at node i is

$$\sum_{k=1}^{\infty} E_k[1-p_{i+1}(m_{i+1})][p_{i+1}(m_{i+1})]^{k-1},$$

where E_k is an Erlang distribution with k phases, each with mean $1/\mu_i$. The above distribution is equivalent to an exponential distribution with parameter $\mu_i[1-p_{i+1}(m_{i+1})]$. Given λ, the arrival rate at the first node, we can apply the above approximation procedure starting from the last node and working backwards. Each node i, i=2,...,M, is studied as an

$M/M/1/m_i$ and node 1 as an M/M/1. Using this algorithm the maximum throughput can be approximated as corresponding to an input value λ^* for which $\lambda^*=\mu_1[1-p_2(m_2)]$.

Pollock, Birge and Alden (1985) described an approximation algorithm for analyzing tandem configurations under the blocking-after-service mechanism assuming exponential and general service times. The approximation algorithm can be used to obtain the queue-length distribution at each node of the system. The algorithm utilizes arguments similar to those used in the Perros and Altiok (1986) algorithm. Mitra and Mitrani (1990) and (1991) introduced a tandem configuration motivated by a *kanban* scheme. The first node was assumed to be saturated and the service times were assumed exponential. The blocking mechanism is as follows. Server i continues to serve customers during the time that node i+1 is full. The served customers remain in node i. Server i gets blocked when the ith node is full and all customers have been served. The authors analyzed approximately the queueing network using a single-node decomposition, and then combining the isolated nodes through a set of fixed-point equations. Massey (1993) obtained the exact joint distribution for a tandem configuration with no intermediate queue, and no queue in front of the first server, i.e. $m_i=1$, $i=1,2,...,M$. Each server provides an exponentially distributed service with the same service rate, equal to the rate of arrivals at the first node. Blocking-before-service was assumed. Bocharov and Rokhas (1980) studied a tandem configuration with exponential service times and no intermediate buffers where the first node was assumed to have an infinite capacity. The authors obtained the maximum throughput using the fact that the rate diagram of the system has a block tri-diagonal rate structure. Also, Bocharov and Spesivov (1986) considered a tandem configuration with no intermediate buffers, where the first node has a finite capacity. The distribution of the inter-arrival time and the service time at the first node were assumed to be phase-type. The service time at the remaining nodes was assumed to be exponentially distributed. External Poisson arrivals occur at each intermediate node. A customer, upon completion of its service at node i, i=2,3,...,M-1, may either proceed to node i+1 or it may be fed back to node i. If a customer attempts to enter node i+1 at a time that the node is full (i.e. busy), the customer is lost. A recursive algorithm is given with which one can obtain various measures such as the probability the queueing network is empty, and the probability the first queue is empty. Makino (1964) and Hunt (1956) studied a queueing network consisting of three servers with exponential service times and no intermediate buffers, assuming that the first node is saturated. They obtained the mean effective service time of the first server. Also, Makino obtained the departure process from the queueing network. A critique of three approximation algorithms was given by Bell (1982). Finally, a unified view of some of the decomposition algorithms reported in the literature and proof of convergence is given by Dallery and Frein (1993).

Tandem configurations with general service times

Choong and Gershwin (1985) proposed an approximation algorithm for analyzing tandem configurations with unreliable servers. The method is similar to the one by Gershwin (1987a) for synchronized tandem configurations. For each server, the service time, the time to fail, and the repair time are assumed exponentially distributed. The same problem was also tackled by Gershwin (1989) and by Hong, Glassey and Seong (1990) whose approximation algorithm is based on the decomposition technique proposed by Gershwin (1987a). Altiok (1989) proposed an approximation algorithm for analyzing tandem configurations with phase-type services. The algorithm is similar to the one by Perros and Altiok (1986). Hildebrand (1968) considered a tandem configuration with saturated first queue, general service times, and blocking-after-service. The author showed formally that the system's throughput is the reciprocal of the effective service time at the first node. He obtained an expression for the system's throughput in some cases. Knot (1970) obtained an approximation for the throughput of a tandem configuration with saturated first queue, under a variety of different service time distributions. Jassim (1987) analyzed approximately a tandem like configuration motivated by an open double loop fault tolerant ring network with finite storage capacities. Muth (1979) considered a tandem configuration with saturated first queue, general service time distributions, and blocking-after-service. He showed that the throughput of this system remains invariant under reversal of the queueing network, i.e. when customers pass through the network in the reverse order beginning with the Mth server and ending at server 1. This was shown by treating each position in a queue as a server with zero service. Similar results were also obtained by Dattatreya (1978).

Also, of interest is the paper by Yamazaki, Kawashima and Sakasegawa (1985). Kelly (1982) considered a tandem configuration in which a customer receives the same service at each node. The service time of a customer is drawn from a general distribution. The first queue is always saturated, blocking-after-service or blocking-before-service may be assumed, and all nodes have the same finite capacity. The author obtains an asymptotic expression for the throughput of the system as the number of nodes increases. Also, in Kelly (1984) the author investigates the effect of re-ordering the sequence of customers on the system' throughput for a similar tandem configuration with no intermediate buffers. The problem of allocating buffers so that to maximize a performance measure has been studied by Altiok and Stidham (1983), Ziedins (1989), MacGregor Smith and Daskalaki (1988), and McClain and Moodie (1991). For further references on this problem see below under synchronized servers. Also, the problem of assigning extra servers to stations in a tandem queueing network has been considered by Hillier and So (1989) and by Lee, Tcha and Yamazaki (1990). Finally, Onvural and Perros (1990) reported on an approximation algorithm for the analysis of a multi-class tandem open queueing network with Coxian service times, Coxian arrival processes, and finite buffers. The approximation algorithm was an extension of the algorithms by Altiok and Ranjan (1988) and Gün and Makowski (1989).

Tandem configuration with constant service times

Newell (1979) analyzed a tandem configuration with constant service times assuming that the first node is saturated. He obtained an explicit expression for the cumulative number of units to depart from server i, i=1,2,..,M, by time t, assuming initially an empty system.

Synchronized servers

The transfer line model discussed in section 4.5 was first analyzed by Buzacott (1967). The effect of station breakdowns was considered in Buzacott (1972), and a review and comparison of some models proposed for transfer lines was given in Buzacott and Hanifin (1978). Gershwin (1987b) extended his method to the case where the servers have different processing speeds. Liu and Buzacott (1989) described a two-server decomposition algorithm for analyzing approximately a generalization of the transfer line model. Jafari and Shanthikumar (1984) also developed a two-server decomposition algorithm for the transfer line model. Ohmi (1981) obtained approximately the throughput of a transfer line assuming a) only one server may be broken down at any time, b) if a server breaks down, then the contents of the buffer in front of it is assumed to be equal to the mean number of customers in the buffer. Yamashita et al (1989) analyzed a transfer line which operated according to demand. Finally, the problem of allocating buffers so that to maximize a performance measure has been studied by Soyster and Toof (1976), Soyster, Schmidt and Rohrer (1979), Sheskin (1976), Jafari and Shanthikumar (1985).

The tandem configuration of synchronized servers can be seen as a discrete model of production systems consisting of machines separated by finite capacity buffers. An alternative way to model such a production system is through a continuous model. This model was proposed independently by Sevast'yanov (1962) assuming time dependent failures, and by Zimmern (1956) assuming operation dependent failures. In this model, the flow of discrete parts in a production system is approximated by a continuous flow. When a server is operational but not idle or blocked, it transfers material to the downstream buffer in a continuous way at a constant rate. The time between failures and the time until repairs are assumed to be exponentially distributed. A continuous flow model was analyzed by Dallery, David and Xie (1989). See also Xie (1989). De Koster (1987), (1988), and De Koster and Wijngard (1989) considered a continuous model in which if the server's buffer becomes empty, the server adjust its speed to the speed of the upstream server. If the upstream server is down, the server is forced to stop. If the downstream buffer is full, the server either adjusts its speed to the downstream server's speed or it stops. The tandem configuration is decomposed into subsystems, each consisting of two successive servers and the intermediate buffer. Each subsystem is analyzed using differential equations. Wesfreid (1985) also used differential equations to analyze a similar tandem configuration consisting of three nodes.

A review of models and analytical results for manufacturing flow line systems can be found in Dallery and Gershwin (1992).

Bibliography

Altiok, T., Approximate analysis of exponential tandem queues with blocking, *Eur. J. Oper. Res.* **11** (1982) 390-398.

Altiok, T., Approximate analysis of queues in series with phase-type service times and blocking, *Oper. Res.* **37** (1989) 601-610.

Altiok, T. and Kao, Z., Bounds for throughput in production/inventory systems in series with deterministic processing times, *IIE Trans.* **21** (1989) 82-85.

Altiok, T. and Ranjan, R., Analysis of production lines with general service times and finite buffers: A two-node decomposition approach, *Engineering Costs and Production Economics* **17** (1988) 155-165.

Altiok, T. and Stidham, S., The allocation of interstage buffer capacities in production lines, *IIE Trans.* **15** (1983) 292-299.

Avi-Itzhak, B., A sequence of service stations with arbitrary input and regular service times, *Mange. Sci.* **11** (1965) 565-571.

Avi-Itzhak, B. and Levy, H., *A sequence of servers with arbitrary input and regular service times revisited*, Tech. Rept. 5-93, Rutcor Research Center, Rutgers Univ., 1993.

Bell, P. C., The use of decomposition techniques for the analysis of open restricted queueing networks, *Oper. Res. Lett.* **1** (1982) 230-235.

Bocharov, P. P. and Rokhas, G. P., On an exponential queueing system in series with blocking, *Prob. Control Infor. Theory* **9** (1980) 441-455.

Bocharov, P. P. and Spesivov, S. S., On multistage queueing system with losses and arriving flow of phase type, *Prob. Control and Infor. Theory* **15** (1986) 465-474.

Brandwajn, A., An iterative solution of two-dimensional birth and death processes, *Oper. Res.* **27** (1979) 595-605.

Brandwajn, A. and Jow, Y.-L. L., Tandem exponential queues with finite buffers, in: *Proc. Int. Seminar on Computer Networking and Performance Evaluation*, Tokyo, Sept 18-20, 1985.

Brandwajn, A. and Jow, Y.-L. L., An approximation method for tandem queues with blocking, *Oper. Res.* **36** (1988) 73-83.

Brandwajn, A. and Sahai, A. K., A note on approximate iterative solution of open tandem networks with blocking, *Perform. Eval.* **10** (1989) 211-218.

Buzacott, J. A., Automatic transfer lines with buffer stocks, *Int. J. Prod. Res.* **5** (1967) 183-200.

Buzacott, J. A., The effect of station breakdowns and random processing times on the capacity of flow lines with in-process storage, *AIIE Trans.* **4** (1972) 308-312.

Buzacott, J. A. and Hanifin, L. E., Models of automated transfer lines with inventory banks - a review and comparison, *AIIE Trans.* **10** (1978) 197-207.

Caseau, P. and Pujolle, G., Throughput capacity of a sequence of queues with blocking due to finite waiting room, *IEEE Trans. Soft. Eng.* **5** (1979) 631-642.

Choong, Y. F. and Gershwin, S. B., A decomposition method for the approximate evaluation of capacitated transfer lines with unreliable machines and random processing times, *IIE Trans.* **19** (1985) 150-159.

Dallery, Y., David, R. and Xie, X.-L., An efficient algorithm for analysis of transfer lines with unreliable machines and finite buffers, *IIE Trans.* **20** (1988) 280-283.

Dallery, Y., David, R. and Xie, X.-L., Approximate analysis of transfer lines with unreliable machines and finite buffers, *IEEE Trans. Aut. Control* **34** (1989) 943-953.

Dallery, Y. and Frein, Y., On decomposition methods for tandem queueing networks with blocking, *Oper. Res.* **41** (1993) 386-399.

Dallery, Y. and Gershwin, S. B., Manufacturing flow line systems: A review of models and analytical results, *Queueing Systems J.* **12** (1992) 3-94.

Dattatreya, E.S., *Tandem queueing systems with blocking*, Ph.D. Thesis, University of California, Berkeley, 1978.

Gershwin, S. B., An efficient decomposition method for the approximate evaluation of tandem queues with finite storage space and blocking, *Oper. Res.* **35** (1987a) 291-305.

Gershwin, S. B., Representation and analysis of transfer lines with machines that have different processing rates, *Ann. Oper. Res.* **8** (1987b) 511-530.

Gershwin, S. B., An efficient decomposition algorithm for unreliable tandem queueing systems with finite buffers, in: Perros and Altiok, (Eds.), *Proc First Int. Workshop on Queueing Networks with Blocking*, (North-Holland, Amsterdam, 1989).

Gershwin, S. B. and Shick, I. C., Modelling and analysis of three-stage transfer lines with unreliable machines and finite buffers, *Oper. Res.* **31** (1983) 354-380.

Gün, L. and Makowski, A. M., An approximation method for general tandem queueing systems subject to blocking, in Perros and Altiok, (Eds.), *Proc. First International Workshop on Queueing Networks with Blocking*, (North-Holland, Amsterdam, 1989).

Hildebrand, D. K., On the capacity of tandem server, finite queue, service systems, *Oper. Res.* **16** (1968) 72-82.

Hillier, F. S. and Boling, R. W., Finite queues in series with exponential or Erlang service times - a numerical approach, *Oper. Res.* **15** (1967) 286-303.

Hillier, F. S. and So, K. C., The assignment of extra servers in tandem queueing system with small or no buffers, *Perform. Eval.* **10** (1989) 219-232.

Hong, Y., Glassey, C. R. and Seong, D., The analysis of a production line with unreliable machines and random processing times, *IIE Trans.* **24** (1992) 77-83.

Hunt, G. C., Sequential arrays of waiting lines, *Oper. Res.* **4** (1956) 674-683.

Jafari, M. A. and Shanthikumar, J. G., *Allocation of buffer storages along a multi-stage automatic transfer line*, Tech. Rept. 84-003, Industrial Engineering Dept., Arizona University, 1984.

Jafari, M. A. and Shanthikumar, J. G., *Determination of optimal buffer storage capacities and optimal allocation in multistage automatic transfer lines*, Tech. Rept. 85-011, Industrial Engineering and Operations Research Dept., Syracuse University, 1985.

Jassim, A. R., *Performance modelling of finite storage double loop fault tolerant ring network*, Tech. Rept., CSR, City University, 1987.

Jun, K.-P. and Perros, H. G., An approximate analysis of open tandem queueing networks with blocking and general service times, *Eur. J. Oper. Res.* **46** (1990) 123-135.

Kelly, F. P., The throughput of a series of buffers, *Adv. Appl. Probab.* **14** (1982) 633-653.

Kelly, F. P., Blocking, reordering, and the throughput of a series of servers, *Stoch. Proc. Appl.* **17** (1984) 327-336.

Knott, A. D., The inefficiency of a series of work stations - a simple formula, *Int. J. Prod. Res.* **8** (1970) 109-119.

de Koster, M. B. M., Estimation of line efficiency by aggregation, *Int. J. Prod. Res.* **25** (1987) 615-626.

de Koster, M. B. M., Approximate analysis of production systems, *Eur. J. Oper. Res.* **37** (1988) 214-226.

de Koster, M. B. M. and Wijngard, J., On the use of continuous vs discrete models in production lines with buffers, in Perros and Altiok, (Eds.), *Proc. First Int. Workshop on Queueing Networks with Blocking* , (North-Holland, 1989).

Kramer, W. and Laugenbach-Belz, M., Approximate formulae for the delay in the queueing system GI/G/1, *Congressbook, 8th Inter. Teletraffic Conf.*, Melbourne (1976) 235-1/8.

Lee, W. T., Tcha, D.-W. and Yamazaki, G., *Server assignment for multi-stage production systems with finite buffers*, Tech. Rept., Management Sci. Dept, KAIST, 1990.

Liu, X.-G. and Buzacott, J. A., A balanced local flow technique for queueing networks with blocking, in: Perros and Altiok, (Eds.), *Proc. First International Workshop on Queueing Networks with Blocking*, (North-Holland, Amsterdam, 1989).

MacGregor Smith, J. and Daskalaki, S., Buffer space allocation in automated assembly lines, *Oper. Res.* **36** (1988) 343-358.

Makino, T., On the mean passage time concerning some queueing problems of the tandem type, *J. Oper. Res. Soc. Japan* **7** (1964) 17-47.

Massey, W. A., Balanced queues in series with communication blocking, *Math. Oper. Res.*, to appear, 1993.

McClain, J. O. and Moodie, D. R., A comment on "buffer space allocation in automated assembly lines", *Oper. Res.* **39** (1991) 857-860.

Mitra, D. and Mitrani, I., Analysis of a kanban discipline for cell coordination in production lines I, *Manage. Sci.* **36** (1990) 1548-1566.

Mitra, D. and Mitrani, I., Analysis of a kanban discipline for cell coordination in production lines II: Stochastic demands, *Oper. Res.* **39** (1991) 807-823.

Muth, E. J., The reversibility property of production lines, *Manage. Sci.* **25** (1979) 152-158.

Newell, G. F., *Approximate behavior of tandem queues*, Lecture Notes in Economics and Mathematical Systems, **171** (Springer, Berlin, 1979).

Ohmi, T., An approximation for the production efficiency of automated transfer lines with in-process storage, *AIIE Trans.* **13** (1981) 22-28.

Onvural, R. O. and Perros, H. G., Approximate analysis of multi-class tandem open queueing networks with Coxian parameters and finite buffers, in: King, Mitrani and Pooley, (Eds.), *Performance '90*, (North-Holland, 1990) 131-140.

Perros, H. G. and Altiok, T., Approximate analysis of open networks of queues with blocking: tandem configurations, *IEEE Trans. Soft. Eng.* **12** (1986) 450-461.

Pollock, S. M., Birge, J. R., and Alden, J. M., *Approximation analysis for open tandem queues with blocking: exponential and general distributions*, IOE Rept. 85-30, Univ. of Michigan, 1985.

Schweitzer, P. J., Aggregate modelling of queueing networks, in: Bonati, (Ed.), *Teletraffic Science for new cost-effective systems*, (North-Holland, 1989), 1522-1528.

Sevast'yanov, B. A., Influence of storage bin capacity on the average standstill time of a production line, *Teoriya Veroyatnostey i ee Primeneniya* **7** (1962) 429-438 (English translation: *Theor. Probab. Appl.* **7** (1962) 429-438).

Shanthikumar, J. G. and Jafari, M. A., *Bounding the performance of tandem queues with finite buffer space*, Tech. Rept., School of Bus. Admin., Univ. of California at Berkeley, 1987.

Sheskin, T. J., Allocation of interstage storage along an automatic production line, *AIIE Trans.* **8** (1976) 146-152.

Soyster, A. L., Schmidt, J. W. and Rohrer, M. W., Allocation of buffer capacities for a class of fixed cycle production lines, *AIIE Trans.* **11** (1979) 140-146.

Soyster, A. I. and Toof, D. I., Some comparative and design aspects of fixed cycle production systems, *Naval Res. Logist. Quart.* **23** (1976) 437-454.

Wesfreid, E., *Dynamic behaviour of N-stage transfer lines with unreliable machines and finite buffers*, Tech. Rept. 456, INRIA, 1985.

Xie, X., *Performance analysis of a transfer line with unreliable machines and finite buffers*, Tech. Rept. 1089, INRIA, 1989.

Yamashita, H., Nakazawa, K., Nagasuka, H., and Suzuki, S., Modeling and analysis of a production line operated according to demand, *JSME* **32** (1989) 491-1497.

Yamazaki, G., Kawashima, T., and Sakasegawa, H., Reversibility of tandem blocking queueing systems, *Manage. Sci.* **31** (1985) 78-83.

Yao, D. and Buzacott, J. A., Queueing models for a flexible machining station part ii: the method of Coxian phases, *Eur. J. Oper. Res.* **19** (1984) 241-252.

Ziedins, I., *Tandem queues with correlated service times and finite capacity*, Tech. Rept., Dept. of Actuarial Mathematics and Statistics, Heriot-Watt University, 1989.

Zimmern, B., Etude de la propagation des arrêts aléatoires dans les chaînes de production, *Revue de Statistique Appliquées* **4** (1956) 85-104.

CHAPTER 5

Approximate Analysis of Arbitrarily Linked Open Queueing Networks With Blocking

5.1. Introduction

In this Chapter, we present two single-node decomposition algorithms for analyzing approximately arbitrary configurations of queueing networks with blocking. The first algorithm, presented in the following section, was developed for feed-forward configurations of open queueing networks with blocking and it is an extension of algorithm 2 described in Chapter 4. The second algorithm, presented in section 5.3, is applicable to arbitrarily linked open queueing networks with blocking. In this approximation algorithm, each node is analyzed in isolation using the maximum entropy methodology. Relevant references are given in section 5.4.

5.2. A single-node decomposition algorithm for feed-forward configurations

A feed-forward configuration is a queueing network in which the nodes are linked arbitrarily without forming loops. An example of such a configuration is given in figure 5.1. The approximation algorithm given in this section is an extension of the single-node decomposition algorithm 2 that was presented in Chapter 4. All service times are assumed to be exponentially distributed and external arrivals are assumed to occur in a Poisson fashion. Customers in each node are served in a FIFO manner. The node to which all external arrivals occur can have a finite or an infinite capacity. The capacity of all other nodes is finite. (External arrivals may occur to more than one node, though for simplicity it has been assumed that all external customers join node 1.) Blocking-after-service is assumed.

The approximation algorithm decomposes the queueing network into individual nodes, each with modified arrival process, service process, and buffer capacity. Each node is then studied in isolation. The parameters of each node are modified as follows: a) the arrival process to each node is approximated by a Poisson distribution; b) the original service mechanism of any node that may become blocked is augmented by a phase-type distribution which represents the blocking delay a customer may experience; and c) the capacity of each finite capacity node is augmented by as many positions as the number of upstream nodes directly linked to it. The phase-type distribution representing the blocking delay a customer may experience upon service completion at a node has a very simple structure. It is concise and consistent from node to node, thus making implementation easy.

Below, we describe the approximation algorithm as it would be applied to the four-node feed-forward configuration shown in figure 5.1. Let λ_1 be the rate of arrivals to node 1, and μ_i be the service rate at node i, i=1,2,3,4. Let r_{ij} be the branching probability that a customer upon completion of its service at node i chooses destination node j. When j=0 the customer departs from the network. We assume $\sum_j r_{ij}=1$, i=1,2,3,4. As before, we set m_i to denote the capacity of the ith node. m_2, m_3 and m_4 are assumed to be finite, and m_1 may be finite or infinite.

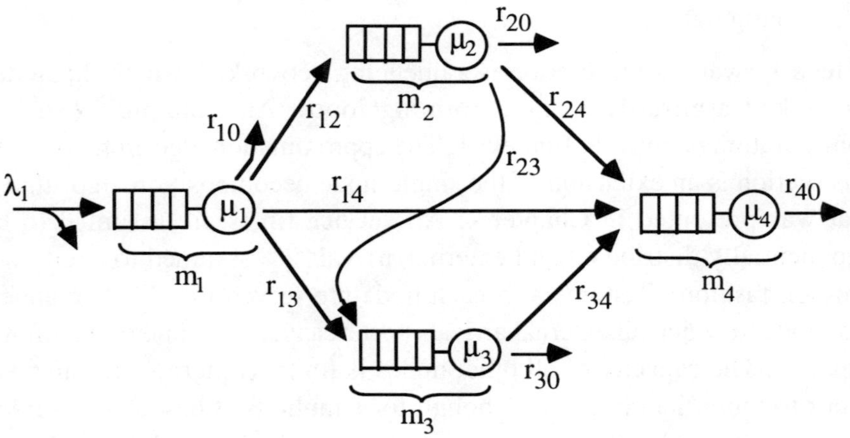

Figure 5.1: A four-node feed-forward configuration.

Finally, let $p_i(n_i)$ be the steady-state probability that there are n_i customers in node i, and let $\pi_{ij}(n)$ be the conditional probability that there are n_j customers in node j upon service completion at server i. In the discussion below we assume that the capacity of node 1, to which all external arrivals occur, is infinite. In the case where the first node has a finite capacity, the algorithm is modified in the same way as algorithm 2 in Chapter 4.

The algorithm begins with the analysis of node 4. Because of the network topology, node 4 cannot get blocked. Since nodes 1, 2 and 3 may send customers to node 4, the capacity of node 4 is increased by 3 and node 4 is analyzed as an M/M/1/m_4+3 node. The equilibrium queue length distribution can be obtained using the familiar M/M/1/N formula once the overall arrival rate λ_4 is known. Assuming that the parameters of the network have been chosen so that node 1 is stable, the effective rate at which customers enter node 4 is $\bar{\lambda}_4=\lambda_1(r_{14}+r_{12}r_{24}+r_{12}r_{23}r_{34}+r_{13}r_{34})$. The overall arrival rate λ_4 can be obtained as a fixed-point problem using the expression $\lambda_4[1-p_4(m_4+3)]=\bar{\lambda}_4$ where $p_4(m_4+3)$ is a function of λ_4.

Before we analyze the remaining nodes, we calculate the blocking delay that a customer at node i, i=1,2,3, experiences if upon service completion it chooses to join node 4. This blocking delay is given by the phase-type structure shown in figure 5.2 We assume that a customer entering this phase-type structure has already received service at node i, i=1,2,3, and has selected node 4 as its destination. With probability $1-\pi_{i4}(m_4)-\pi_{i4}(m_4+1)-\pi_{i4}(m_4+2)$ the customer will immediately enter node 4. This event corresponds to the upper branch in figure 5.2. The other paths represent the event that the customer finds the buffer full. If no other customers are waiting for buffer space, then the customer from node i will be forced to wait at node i until a service completion occurs at node 4. This time has an exponential distribution with parameter μ_4. This event corresponds to the second branch from the top in figure 5.2. If a customer arrives from node i to find 1 customer already waiting for buffer space, then the customer will get blocked at node i for a period of time which is exponentially distributed with parameter μ_4. This event corresponds to the third from the top branch in figure 5.2. The bottom branch corresponds to the event where a customer from node i finds 2 customers waiting for buffer space. The quantities $\pi_{i4}(.)$, i=1,2,3, are determined using Little's relation as follows. Let $\pi_4(n)$ be the probability that any customer arriving at node 4 will see n customers. Then, we have

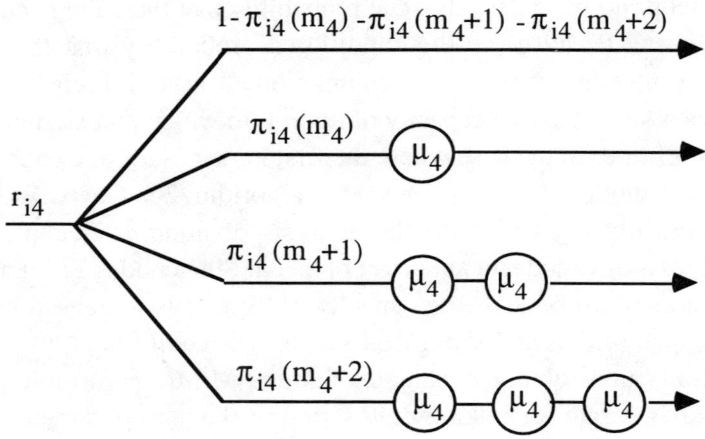

Figure 5.2: The blocking delay at node 4.

$$\bar{\lambda}_4\pi_4(m_4)/\mu_4 = p_4(m_4+1)$$

$$\bar{\lambda}_4\pi_4(m_4+1)/\mu_4 = p_4(m_4+2)$$

$$\bar{\lambda}_4\pi_4(m_4+2)/\mu_4 = p_4(m_4+3).$$

From the above expressions we can calculate $\pi_4(.)$, since we have already obtained the queue-length distribution at node 4. Next, we estimate $\pi_{i4}(.)$, i=1,2,3, by simply allocating $\pi_4(.)$ proportionally over the three upstream nodes. We have

$$\pi_{14}(m_4+i) = r_{14}\,\pi_4(m_4+i)/r$$

$$\pi_{24}(m_4+i) = r_{12}r_{24}\,\pi_4(m_4+i)/r$$

$$\pi_{34}(m_4+i) = (r_{13}r_{34}+r_{12}r_{23}r_{34})\,\pi_4(m_4+i)/r,$$

for i=0,1,2, and $r=r_{14}+r_{12}r_{24}+r_{13}r_{34}+r_{12}r_{23}r_{34}$. Once all the parameters of the phase-type distribution in figure 5.2 are determined, the distribution is approximated by a C_2 distribution with parameters v_{i41}, v_{i42}, a_{i4}, i = 1,2,3, using the method of moments.

The third node is now considered in isolation. The capacity of node 3 is increased by two since both nodes 1 and 2 may be blocked by node 3. The service mechanism of node 3 is augmented, as shown in figure 5.3, in order to include the blocking delay that a customer may experience if it gets blocked by node 4. This blocking delay is given by the C_2 distribution obtained above. Assuming that the arrival process to node 3 is Poisson, we can analyze this node in isolation as an $M/PH/1/m_3+2$ queue. (In the original implementation of this algorithm, all M/PH/1/m queues were analyzed using the matrix-geometric procedure.) In order to carry out the analysis, we need to know the total arrival rate λ_3 rather than the effective rate $\bar{\lambda}_3$. This quantity can be computed numerically from the expression $\bar{\lambda}_3=\lambda_3[1-p_3(m_3+2)]$, where $\bar{\lambda}_3=\lambda_1(r_{13}+r_{12}r_{23})$, as follows. Assuming an initial value for λ_3, we solve numerically to obtain the queue-length distribution of node 3. Using the value for $p_3(m_3+2)$ we calculate a new value for λ_3, and then we go back and re-calculate the queue-length distribution. We continue to iterate in this fashion until the successive values for λ_3 converge. This fixed-point problem is similar to the one employed in algorithms 1 and 2 in Chapter 4.

Before we proceed with the analysis of node 2 we construct two C_2 distributions related to node 3. This is done so that the solution of each node in isolation is both simple and consistent from node to node. First, the representation of the augmented service mechanism at node 3, shown in figure 5.3, is condensed to a two-phase Coxian distribution with parameters v_{31}, v_{32} and a_3. This C_2 distribution represents the effective service time at node 3. We use this distribution to construct the blocking delay a customer at node i, i=1,2, may experience, if it chooses to join node 3 upon completion of its service. This blocking delay is shown in figure 5.4. The probabilities $\pi_{i3}(.)$, i=1,2, are obtained in the same way as in node 4. This phase-type distribution is approximated by a two-phase Coxian distribution with parameters v_{i31}, v_{i32} and a_{i3}, i=1,2, using the method of moments. Node 2 can now be analyzed following similar

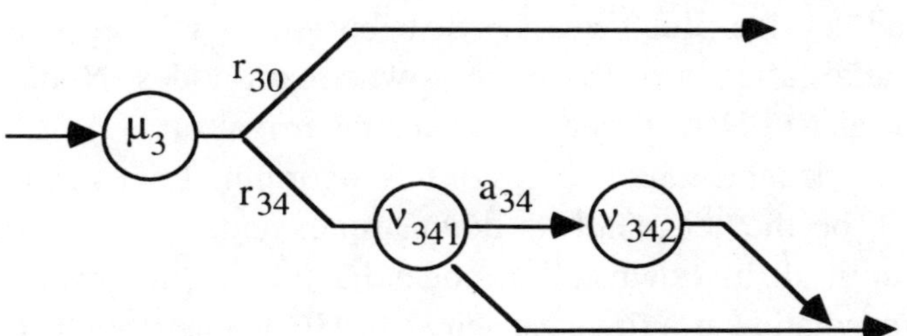

Figure 5.3: Effective service time at node 3.

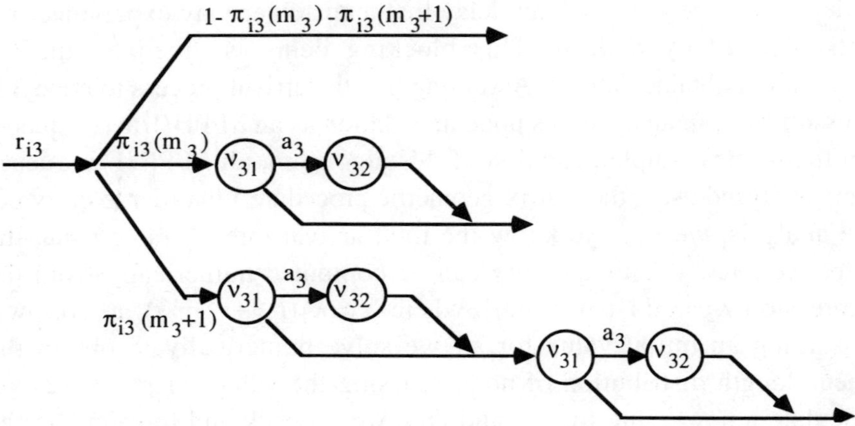

Figure 5.4: The blocking delay at node 3.

arguments as in node 3. In particular, its service time is augmented so that in addition to its regular service, it comprises of two C_2 distributions with parameters v_{231}, v_{232}, a_{23}, and v_{241}, v_{242}, a_{24}, reflecting the blocking delay at node 3 and 4 respectively. Figure 5.5 gives the phase-type distribution representing node 2's effective service time. Assuming that the arrival process to node 2 is Poisson, we can analyze this node in isolation as an M/PH/1/m_2+1 queue. As in node 3, we need to know the total arrival rate λ_2 rather than the effective rate $\bar{\lambda}_2$. This quantity can be computed numerically as a fixed point problem using the expression $\bar{\lambda}_2 = \lambda_2[1-p_2(m_2+1)]$, where $\bar{\lambda}_2 = r_{12}\lambda_1$.

The analysis of node 1 is carried out in the same way. We construct the blocking delay due to node 2 and then collapse it approximately to a C_2 distribution. We do the same for the blocking delays due to nodes 3 and 4. Then, we construct the effective service time at node 1. This consists of the original service plus three C_2 distributions, each representing the blocking delay at each of the three downstream nodes. Node 1 is then analyzed as an M/PH/1/m_1 queue with a total arrival λ_1.

We summarize the main features of this algorithm. Let us consider node i and let D_i be the set of all its downstream nodes. Node i cannot be analyzed until all its downstream nodes have been analyzed. Node i is analyzed in isolation as a finite capacity M/PH/1/N queue with an effective service time consisting of the original service time plus one C_2 distribution per downstream node j, $j \in D_i$. This C_2 distribution reflects the blocking delay due to the finite capacity of node j and it is obtained as follows. After node j is analyzed in isolation, its effective service time is collapsed into a C_2 distribution using the three moment approximation. This C_2 distribution is then used to construct a phase-type distribution of the blocking delay a customer in an upstream node may experience. This phase-type distribution is finally collapsed into a C_2 distribution, using the three moment approximation, which is used in the effective service time of node i.

The algorithm yields the queue-length distribution of each node. Validation tests of this approximation algorithm showed that it has a good accuracy. The average relative error in the queue-length distribution and in the mean queue-length of each node was observed to be about 5%. As in the case of tandem configurations, in the case of large buffers the relative error in the queue-length distribution gets higher towards the tail of the distribution.

We note that in order to calculate the probabilities $\pi_{ij}(.)$ we concatenate all the three arrival streams into a single stream. This approach works well when the arrival streams are balanced. If the arrival streams are not balanced, one can use weights to concatenate the streams. Alternatively, the node can be analyzed in isolation by explicitly representing each of the

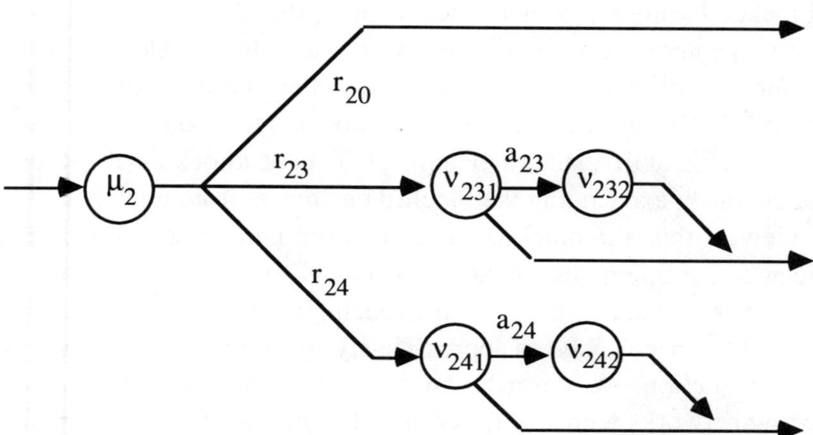

Figure 5.5: The effective service time of node 2.

streams in the Markov process formulation. This of course will increase the complexity of the solution algorithm, but the accuracy will be improved.

Finally, in this algorithm we approximate various phase-type distributions by two-phase Coxian distributions using the method of moments. This method can be applied when the c^2 of a phase-type distribution is greater than one. In the original implementation of this algorithm, the two moment approximation, described in section 1.5, was used when the c^2 of a phase-type distribution was less than one.

5.3. Arbitrary configurations

Let us consider an open queueing network consisting of arbitrarily linked finite capacity queues with blocking-after-service. Deadlock may occur due to the arbitrary interconnection of the nodes. For example, let us suppose that node i is full and it is blocked by node j. A deadlock will occur if the customer in service at node j chooses to go to node i upon completion of its service. We assume that a deadlock is detected instantaneously and resolved without any delay by exchanging all the blocked customers simultaneously. That is, in the above example, the blocked customers at nodes i and j will be exchanged simultaneously. In general, this simple scheme for resolving deadlocks may violate the first-blocked-first-enter priority rule associated with the blocking-after-service mechanism (see Chapter 1). For instance, let us assume that node i and k are blocked by node j in that order. That is, the blocked customer from node i will enter node j first if a departure occurs from node j. Let us now consider the following situation. Suppose that the departing customer from node j chooses node k as its destination and that node k is full at that moment. This will cause a deadlock. The deadlock is resolved by simultaneously exchanging the blocked customers from node k and node j. In view of this, the blocked customer from node k enters node j first while node i still remains blocked by node j. Thus, the first-blocked-first-enter rule is violated! This type of queueing network with blocking and deadlocks has been analyzed approximately (see references, section 5.4).

In this section, we describe an approximation algorithm for open queueing networks with arbitrarily linked finite capacity nodes assuming repetitive-service blocking with random destination. That is, if a customer at node i finds that its selected destination node is full at the moment it completes its service, the customer receives another service at node i. It then attempts to enter a new destination node chosen independently of the destination node selected the last time. This is repeated until the customer successfully enters a destination node. We note that deadlocks do not arise due to the nature of the blocking mechanism. The algorithm presented below is a single-node decomposition algorithm where each individual node is analyzed using the maximum entropy methodology.

A maximum entropy algorithm

Let us consider an open queueing network of arbitrarily linked finite capacity nodes. There is an external arrival process to each node. All service times and inter-arrival times are assumed to have a generalized exponential (GE) distribution (see section 1.4 in Chapter 1). In a GE distribution, a customer with probability r receives an exponential service time with parameter ra, or with probability 1-r it receives a zero service time. This means that several successive customers may receive zero service. In view of this, a GE service time distribution is equivalent to an exponential service time with geometrically distributed bulk departures. Similarly, if the inter-arrival time to a node has a GE distribution, then this is equivalent to saying that the arrival process is Poisson with geometrically distributed bulk arrivals. A single server node with capacity m and GE inter-arrival and service time distributions will be notated as GE/GE/1/m.

The approximation algorithm decomposes the queueing network into individual nodes with modified arrival and service distributions. Each node is then analyzed in isolation using the principle of maximum entropy. Before we present the approximation algorithm, we divert to discuss the principle of the maximum entropy and how it is used to analyze the GE/GE/1/m queue. The analysis of this single server queue forms the basis of the approximation algorithm.

Shannon (1948) defined entropy as the measure of uncertainty contained in a probability distribution that describes the states of a system. For the computation of the entropy Shannon proposed the entropy functional

$$H(p) = -\sum_{n \in S} p(n)\ln(p(n)) ,$$

where S is the set of all possible states of a system, n is a certain state of the system, and p(n) is the probability of being in state n. The work of Shannon was extended by Jaynes (1957), who proposed an elegant and purely probabilistic treatment of statistical thermodynamics.

Let us consider a system that has a finite or countable set of states S. The prior information regarding the system is described (without loss of generality) in the form of the following I+1 constraints

$$\sum_{n \in S} p(n) = 1, \tag{5.3.1}$$

$$\sum_{n \in S} f_i(n)p(n) = \langle f_i(n) \rangle, \quad 1 \leq i \leq I, \tag{5.3.2}$$

where $\{\langle f_i(n) \rangle\}$ is a set of expectations defined through appropriate functions $\{f_i(n), n \in S\}$ of the system states. In general, since $I+1<|S|$, there is an infinite number of probability distributions $\{p(n)\}$ that satisfy constraints (5.3.1) and (5.3.2). The maximum entropy distribution can be obtained by the method of the Lagrange's undetermined multipliers leading to the solution

$$p(n) = \frac{1}{G} \prod_{i=1}^{I} x_i^{f_i(n)}, \tag{5.3.3}$$

where $x_i = e^{-\theta_i}$ and θ_i, $i=1,2,\ldots,I$, are the Lagrangian multipliers determined from the set of constraints (5.3.2). G is the normalizing constant given by

$$G = e^{-\theta_0} = \sum_{n \in S} \prod_{i=1}^{I} x_i^{f_i(n)},$$

where θ_0 is the Lagrangian multiplier associated with the mormalizing constraint (5.3.1). It can be easily shown that the Lagrangian multipliers satisfy the relations

$$-\frac{\partial \theta_0}{\partial \theta_i} = \langle f_i \rangle, \quad i = 1,2,\ldots,I.$$

Therefore, the maximum entropy functional takes the form

$$\max_p\{H(p)\} = \theta_0 + \prod_{i=1}^{I} \theta_i \langle f_i \rangle .$$

Let us now consider a GE/GE/1/m queue. Let s, $s\lambda$ and r, $r\mu$ be the parameters of the inter-arrival and service time distribution respectively. Customers are lost if they arrive at the time when the node is full. Below we give the steady-state equations of the GE/GE/1/m queue, and subsequently we obtain the maximum entropy solution using the above formalism. Let p(n) be the steady-state probability that there are n customers in the node, n=0,1,...,m. The steady-state equations will be obtained using the bulk interpretation of the GE distribution. We will make use of the following three probabilities.

Pr{Given that the node is empty, a bulk of j>0 customers arrives and departs in zero time}

$$= s\lambda \sum_{j=1}^{\infty} s(1-s)^{j-1}(1-r)^{j}$$

$$= s^2\lambda(1-r) \sum_{j=1}^{\infty} (1-s)^{j-1}(1-r)^{j-1}$$

$$= \frac{s^2\lambda(1-r)}{r(1-s)+s} \; .$$

Pr{Given that the node is empty, a bulk of $j \geq i$ customers arrives and the first j-i receive zero service}

$$= s\lambda \sum_{j=1}^{\infty} r(1-r)^{j-i} s(1-s)^{j-1}$$

$$= \frac{s^2\lambda r(1-s)^{i-1}}{r(1-s)+s} \; .$$

Pr{Given that the node is empty, a bulk of $j \geq m$ customers arrives and at most j-m customers receive zero service}

$$= s\lambda \sum_{j=m}^{\infty} \sum_{i=0}^{j-m} r(1-r)_i s(1-s)_{j-1}$$

$$= \frac{s\lambda r(1-s)_{m-1}}{r(1-s)+s}.$$

The steady-state equations are as follows.
For n=0 we have

$$s\lambda p(0) = \frac{s^2\lambda(1-r)}{r(1-s)+s} p(0) + \sum_{i=1}^{m} r\mu(1-r)^{i-1}p(i),$$

or

$$\frac{r}{r(1-s)+s} s\lambda p(0) = r\mu \sum_{i=1}^{m} (1-r)^{i-1}p(i). \qquad (5.3.4)$$

For n=1,2,...,m-1, we have

$$(s\lambda+r\mu)p(n) =$$

$$\frac{s^2\lambda r(1-s)^{n-1}}{r(1-s)+s} p(0)$$

$$+ s\lambda \sum_{i=1}^{n-1} s(1-s)^{n-1-i}p(i)$$

$$+ r\mu \sum_{i=n+1}^{m} r(1-r)^{i-n-1}p(i). \qquad (5.3.5)$$

Finally, for n=m we have

$$r\mu p(m) = \frac{s\lambda r(1-s)^{m-1}}{r(1-s)+s} p(0) + s\lambda \sum_{i=1}^{m-1} (1-s)^{m-i-1}p(i). \qquad (5.3.6)$$

Typically, one would attempt to solve the above equations in order to obtain p(n), n=0,1,...,m. However, here we use a different approach. Using the maximum entropy principle we obtain a solution for p(n). As it will be seen, this solution contains three unknowns. These unknowns are then calculated by substituting the maximum entropy expression into the above steady-state equations.

The maximum entropy solution for p(n), n=0,1,...,m, is obtained by maximizing the function

$$H(p) = - \sum_{n=0}^{m} p(n)\log p(n)$$

subject to constraints (5.3.7) to (5.3.10) given below.

1. Normalization

$$\sum_{n=0}^{m} p(n) = 1. \qquad (5.3.7)$$

2. The percentage of time the node is utilized is equal to ρ, where $\rho=\bar{\lambda}/\mu$ and $\bar{\lambda}$ is the effective arrival rate. That is

$$\sum_{n=1}^{m} p(n)=\rho.$$

This constraint is written out into the convenient form

$$\sum_{n=0}^{m} \zeta_0(n)\, p(n) = \rho, \text{ where } \zeta_0(n) = \min\{1, \max(0,n)\}. \qquad (5.3.8)$$

3. The mean number of customers in the node is equal to L. That is

$$\sum_{n=0}^{m} np(n) = L. \qquad (5.3.9)$$

4. The probability that the node is full is f. That is p(m)=f. This constraint is written out into the convenient form

$$\sum_{n=0}^{m} \zeta_1(n)p(n) = f, \text{ where } \zeta_1(n) = \max\{0, n-m+1\}. \qquad (5.3.10)$$

Assuming now that we know ρ, L and f, we can use Lagrange's method of undetermined multipliers to obtain p(n) that maximizes H(p) subject to the above four constraints. From (5.3.3) we have

$$p(n) = p(0)\, x^{\zeta_0(n)}\, y^n\, z^{\zeta_1(n)}, \qquad (5.3.11)$$

where $x=e^{-\theta_1}$, $y=e^{-\theta_2}$, $z=e^{-\theta_3}$, and θ_1, θ_2 and θ_3 are the Langrangian multipliers corresponding to constraints (5.3.8), (5.3.9) and (5.3.10). From (5.3.11) we obtain the following simple recursion:

$$\begin{cases} p(m) = zyp(m-1) \\ p(n) = yp(n-1), \ 1 \le n \le m-1 \\ p(1) = xyp(0). \end{cases} \tag{5.3.12}$$

The quantities x, y and z can be calculated by substituting (5.3.11) into the steady-state equations (5.3.4) to (5.3.6). We have

$$x = \frac{sr\rho}{s\rho(1-r)+r} \tag{5.3.13a}$$

$$y = \frac{s\rho+r(1-s)}{s\rho(1-r)+r} \tag{5.3.13b}$$

$$z = \frac{1}{1-(1-r)y} \, . \tag{5.3.13c}$$

We note that x and y have been shown to be invariant to the node capacity. That is, they also hold for the GE/GE/1 queue.

We conclude the analysis of the GE/GE/1/m queue with the calculation of the probability that an arriving customer will find the queue full and it will get lost. This probability will be used in the approximation algorithm presented below. We first note that since bulks arrive in a Poisson fashion with parameter $s\lambda$, the queue-length distribution of the number of customers in the node as seen by a bulk is the same as the time average distribution p(n), n=0,1,..,m, obtained above. Let us consider a tagged customer in the arriving bulk. Depending upon the position of the customer in the bulk and how many customers in the node the bulk sees, we distinguish the following mutually exclusive events:

1. The bulk arrives at the node and sees m customers. In this case, the tagged customer gets lost irrespective of its position in the bulk. This event occurs with probability p(m).
2. The bulk containing the tagged customer sees n customers in the node, $1 \le n \le m-1$. The bulk size is at least m-n+1 and the tagged customer is

one of the customers in the bulk that gets lost. This event occurs with probability

$$\sum_{n=1}^{m-1} p(n) \sum_{j=m-n+1}^{\infty} \frac{js(1-s)^{j-1}}{1/s} \frac{j-(m-n)}{j} = \sum_{n=1}^{m-1} p(n)(1-s)^{m-n},$$

where the term $js^2(1-s)^{j-1}$ is the probability that the bulk containing the tagged customer is of size j, and the term (j-(m-n))/j is the probability that the tagged customer occupies one of the positions m-n+1, m-n+2,...,j in the bulk.

3. The bulk sees zero customers in the node. The bulk size j is at least m+1, of which at most j-(m+1) customers depart with zero service. The next m customers of the bulk occupy the node and the remaining customers, of which the tagged customer is one, get lost. This event occurs with probability

$$p(0) \sum_{j=m+1}^{\infty} \frac{js(1-s)^{j-1}}{1/s} \sum_{i=0}^{j-m-1} r(1-r)^i \frac{j-m-i}{j} = p(0) \frac{r(1-s)^m}{r(1-s)+s}.$$

Thus, the probability that an arriving customer finds the node full and gets lost is given by the expression

$$p(0) \frac{r(1-s)^m}{r(1-s)+s} + \sum_{n=1}^{m-1} p(n)(1-s)^{m-n} + p(m). \qquad (5.3.14)$$

Let us now consider an open queueing network consisting of M arbitrarily linked nodes. Each node i has a finite capacity m_i and a GE service time. There is an external arrival process for each queue. Repetitive-service blocking with random destination is assumed. The state of the queueing network can be described by the vector $\mathbf{n}=(n_1,n_2,...,n_M)$ where n_i is the number of customers in node i. Let $p(\mathbf{n})$ be the steady-state probability that the queueing network is in state $\mathbf{n}$, and let $p_i(n_i)$ be the marginal probability that there are n_i customers in the ith node. The form of $p(\mathbf{n})$ can be specified by maximizing the entropy function

$$H(p) = - \sum_{\mathbf{n}} p(\mathbf{n}) \log p(\mathbf{n}),$$

subject to the normalization constraint $\sum p(\mathbf{n})=1$, and the marginal constraints of the utilization ρ_i of node i, mean number of customers L_i in node i, and probability f_i that node i is full. That is

$$\sum_{n_i=0}^{m_i} \zeta_0(n_i)p_i(n_i) = \rho_i,\ \zeta_0(n_i)=\min\{1,\max(0,n_i)\}, \qquad (5.3.15)$$

$$\sum_{n_i=0}^{m_i} n_i p_i(n_i) = L_i, \qquad (5.3.16)$$

$$\sum_{n_i=0}^{m_i} \zeta_i(n_i)p_i(n_i) = f_i,\ \zeta_i(n_i)=\max\{0,n_i-m_i+1\}, \qquad (5.3.17)$$

where i=1,2,..,M. (These constraints are similar to the constraints (5.3.8) to (5.3.10).) Using Lagrange's method of undetermined multipliers we have

$$p(\mathbf{n}) = \frac{1}{G}\prod_{i=1}^{M} x_i^{\zeta_0(n_i)}\, y_i^{n_i}\, z_i^{\zeta_i(n_i)}, \qquad (5.3.18)$$

where G is the normalizing constant, $x_i=e^{-\gamma_{i1}}$, $y_i=e^{-\gamma_{i2}}$, $z_i=e^{-\gamma_{i3}}$, and γ_{i1}, γ_{i2}, and γ_{i3} are the Langrangian multipliers for node i corresponding to constraints (5.3.15), (5.3.16) and (5.3.17). From (5.3.18) we have

$$G = \sum_{\mathbf{n}} \prod_{i=1}^{M} x_i^{\zeta_0(n_i)}\, y_i^{n_i}\, z_i^{\zeta_i(n_i)}$$

$$= \prod_{i=1}^{M} \sum_{n_i=0}^{m_i} x_i^{\zeta_0(n_i)}\, y_i^{n_i}\, z_i^{\zeta_i(n_i)}. \qquad (5.3.19)$$

Expression (5.3.18) is not very useful since we do not know x_i, y_i, and z_i, i=1,2,...,M. Unlike the case of the GE/GE/1/m queue, it is not possible to obtain these parameters by substituting (5.3.18) into the steady-state equations of the queueing network. We note, however, that if we use (5.3.19) into (5.3.18), $p(\mathbf{n})$ can be written as

$$p(\mathbf{n}) = \prod_{i=1}^{M} p_i(n_i), \qquad (5.3.20)$$

where

$$p_i(n_i) = \frac{x_i^{\zeta_0(n_i)}\, y_i^{n_i}\, z_i^{\zeta_i(n_i)}}{\sum_{n_i=0}^{m_i} x_i^{\zeta_0(n_i)}\, y_i^{n_i}\, z_i^{\zeta_i(n_i)}} .$$

The above expression is the maximum entropy solution of the marginal probability $p_i(n_i)$ of node i when node i is modelled as a GE/GE/1/m queue. Therefore, expression (5.3.20) suggests that the joint probability $p(\mathbf{n})$ is the product of the maximum entropy solution of each node, where each node is treated as a GE/GE/1/m queue in isolation. This observation leads to a single-node decomposition algorithm where each node is analyzed as a GE/GE/1/m queue. Below we describe in detail how the parameters of the arrival and service processes of each node are characterized so that the node can be analyzed as a GE/GE/1/m queue in isolation.

We first recall that in a GE(r,ra) distribution, r and the squared coefficient of variation c^2, are related to a and mean $1/\mu$ by the expressions: $r=2/(1+c^2)$ and $a=\mu$. Let us now consider node i. We assume that the service time of the node has a $GE(r_i,r_ia_i)$ distribution with $r_i=2/(1+c_{si}^2)$ and $a_i=\mu_i$, where μ_i is the service rate and c_{si}^2 is the squared coefficient of variation of the service time. Let π_{ij} be the probability that a customer upon completion of its service at node i is blocked by node j, and let r_{ij} be the branching probability that a customer upon service completion at node i attempts to join node j, $i \neq j$, $i,j=1,2,\ldots,M$. Then, the probability π_{di} that a customer becomes blocked upon completion of its service at node i is

$$\pi_{di} = \sum_{j=1}^{M} r_{ij}\pi_{ij} .$$

Due to the repetitive-service blocking, the effective service time of a customer is the random sum of the $GE(r_i,r_ia_i)$ distribution over a geometric counter with parameter $1-\pi_{di}$. As it was shown in section 1.4, this random sum has the distribution

$$GE\left(\frac{r_i}{(1-\pi_{di})(1-r_i)+r_i}, \frac{(1-\pi_{di})r_ia_i}{(1-\pi_{di})(1-r_i)+r_i}\right).$$

Therefore, the service rate $\bar{\mu}_i$, and the squared coefficient of variation $\bar{c}^2_{si}$ of the effective service time are

$$\bar{\mu}_i = (1-\pi_{di})\mu_i \tag{5.3.21a}$$

$$\bar{c}^2_{si} = \pi_{di}+c^2_{si}\,(1-\pi_{di}). \tag{5.3.21b}$$

The probabilities π_{ij} are calculated below. We now proceed to characterize the arrival process to node i. We distinguish between the overall arrival process and the effective arrival process. The overall arrival process to node i includes all the arrivals from the upstream nodes and from outside the network that entered the node plus all the arrivals that were blocked because the node was full. The effective arrival process includes only the arrivals from the upstream nodes and from outside the network that actually entered the node. Let λ_i, c^2_i and $\bar{\lambda}_i$, $\bar{c}^2_i$ be the arrival rate and the squared coefficient of variation of the inter-arrival time of the overall arrival process and of the effective arrival process respectively. The following simple relations hold between these quantities:

$$\lambda_i = \frac{\bar{\lambda}_i}{1-\pi_i}, \tag{5.3.22}$$

$$c^2_i = \frac{\bar{c}^2_i - \pi_i}{1-\pi_i}, \tag{5.3.23}$$

where π_i is the probability that a customer arriving at node i (either from an upstream node or from outside the queueing network) finds the node full. (For a more general relation between the overall and the effective arrival processes see section 4.3.) An expression for π_i is given below.

Now, let us consider the various flows into node i. Let node j be an upstream node. Then, the effective arrival rate from node j to node i is $\bar{\lambda}_j r_{ji}$. Using (5.3.22) we have that the overall arrival rate from node j to node i is $\bar{\lambda}_j r_{ji}/(1-\pi_{ji})$. (This arrival rate is equal to the rate at which customers from node j enter node i or are blocked by node i.) In addition to the flows from the upstream nodes there is an external arrival process. Let λ_{0i} and c^2_{0i} be the arrival rate and the squared coefficient of variation of the inter-arrival time of the external process. The overall arrival rates of the flows into node i are shown in figure 5.6.

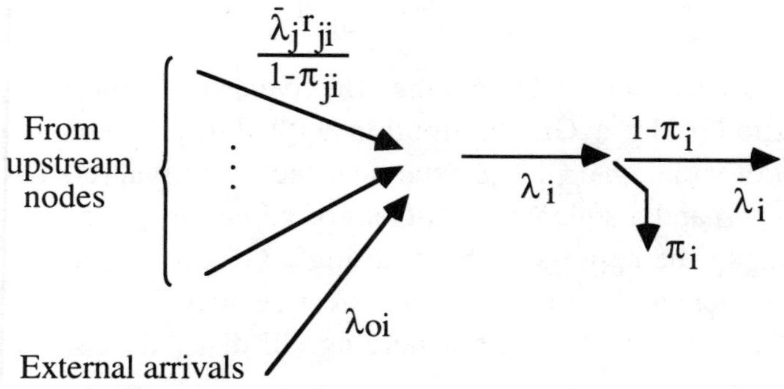

Figure 5.6: The arrival processes to node i.

In view of the above discussion, probability π_i is given by the expression

$$\pi_i = \frac{\lambda_{0i}\pi_{0i} + \sum_{j=1}^{M} \bar{\lambda}_j r_{ji}\pi_{ji}/(1-\pi_{ji})}{\lambda_{0i} + \sum_{j=1}^{M} \bar{\lambda}_j r_{ji}/(1-\pi_{ji})}, \tag{5.3.24}$$

where π_{0i} is the probability that an externally arrived customer finds the node full.

We now proceed to calculate the squared coefficient of variation $\bar{c}_i^2$ of the effective arrival process to node i. Let node j be an upstream node. The inter-departure time from this node is assumed to be a GE distribution with rate $\bar{\lambda}_j$ and squared coefficient of variation $\bar{c}_{dj}^2$. (This inter-departure time does not include those customers which were blocked and were forced to receive another service.) The squared coefficient of variation $\bar{c}_{dj}^2$ is approximated by that of a GE/GE/1 queue. We have

$$\bar{c}_{dj}^2 = \bar{\rho}_j^2\, \bar{c}_{sj}^2 + (1-\bar{\rho}_j)\, \bar{c}_j^2 + (1-\bar{\rho}_j)\bar{\rho}_j, \tag{5.3.25}$$

where $\bar{\rho}_j = \bar{\lambda}_j/\bar{\mu}_j$. Now, part of this departure process becomes the effective arrival process from node j to node i. Seeing that we have assumed that the inter-departure process from node j is a GE distribution, the effective arrival process from node j to node i is also GE with a squared coefficient of variation $\bar{c}_{dji}^2$ given from the expression for splitting a GE distribution (see section 1.4). We have

$$\bar{c}^2_{dji} = 1+r_{ji}\,(\bar{c}^2_{dj} - 1). \tag{5.3.26}$$

So, we now have characterized the effective arrival process from each upstream node j by a GE distribution with rate $\bar{\lambda}_j r_{ji}$ and a squared coefficient of variation $\bar{c}^2_{dji}$. The external effective arrival rate has a mean $\bar{\lambda}_{0i}=\lambda_{0i}(1-\pi_{0i})$ and a squared coefficient of variation $\bar{c}^2_{0i}=\pi_{0i}+(1-\pi_{0i})c^2_{0i}$, obtained using the expression for splitting a GE distribution. Thus, the squared coefficient of variation of the effective arrival process $\bar{c}^2_i$ can be obtained from the expression for merging GE distributions (see section 1.4). We have

$$\bar{c}^2_i = -1 + \frac{1}{\dfrac{\bar{\lambda}_{0i}}{\bar{\lambda}_i}\dfrac{1}{\bar{c}^2_{0i}+1} + \displaystyle\sum_{j=1}^{M}\dfrac{\bar{\lambda}_j r_{ji}}{\bar{\lambda}_i}\dfrac{1}{\bar{c}^2_{dji}+1}}, \tag{5.3.27}$$

Finally, it can be shown that probability π_{ji} is given by the expression

$$\pi_{ji} = (1-s_{ji})^{m_i}\frac{\bar{r}_i}{\bar{r}_i(1-s_{ji})+s_{ji}}\,p_i(0) + \sum_{n=1}^{m_i}(1-s_{ji})^{m_i-n}\,p_i(n), \tag{5.3.28}$$

where $\bar{r}_i=2/(1+\bar{c}^2_{si})$ and

$$s_{ji} = \begin{cases} 2\left[1+\dfrac{\bar{c}^2_{dji}-\pi_{ji}}{1-\pi_{ji}}\right]^{-1} & ,j\neq 0 \\[2ex] \dfrac{2}{1+c^2_{0i}} & ,j=0. \end{cases}$$

The approximation algorithm is summarized below. We note that the service time of node i, i=1,2,...,M, is defined by the service rate μ_i and the squared coefficient of variation c^2_{si}.

Step 0: For i=1,2,...,M, initialize $\bar{c}^2_{di}$. For i=1,2,...,M, with $r_{ij}>0$, remove feedbacks by setting $\mu_i=\mu_i(1-r_{ii})$, $c^2_{si}=r_{ii}+(1-r_{ii})c^2_{si}$, $r_{ij}=r_{ij}/(1-r_{ii})$, $i\neq j$, j=1,2,...,M, and $r_{ii}=0$.

Step 1: Solve the traffic equations $\bar{\lambda}_i=\bar{\lambda}_{0i}+\Sigma_j r_{ji}\,\bar{\lambda}_j$, i=1,2,...,M.

Step 2: Solve the system of non-linear equations (5.3.28) for π_{ij}, i=0,1,...,M, j=1,2,...,M, $i\neq j$, using (5.3.21) to (5.3.27).

Analyze each node i as a GE/GE/1/m queue with arrival parameters λ_i, c_i^2, and service parameters $\bar{\mu}_i$, $\bar{c}_{si}^2$, i=1,2,...,M.

Step 3: Calculate new values for $\bar{c}_{di}^2$ from (5.3.25), for i=1,2,...,M.

Step 4: Check for convergence of the values $\bar{c}_{di}^2$, i=1,2,...,M. Stop if convergence has occurred, else return to step 1.

The algorithm yields the marginal queue-length distribution. Validation tests showed that it has a good accuracy. The average relative error in the queue-length distribution and in the mean queue-length of each node was about 10%. The algorithm gives better results if all queues have finite capacities and there are no loops, i.e. the queueing network is a feed-forward configuration. The results tend to have a high relative error when service times have a high variability and/or servers are highly utilized, particularly when there are loops.

5.4. References

The approximation algorithm for feed-forward configurations, presented in section 5.2 is due to Perros and Snyder (1989). This algorithm is a computationally efficient version of an earlier algorithm by Altiok and Perros (1987). The maximum entropy algorithm presented in section 5.3 is due to Kouvatsos and Xenios (1989a). It has been further improved by Kouvatsos and Denazis (1991). Expression (5.3.25) for the squared coefficient of variation $\bar{c}_{di}^2$, used in the maximum entropy algorithm, is due to Kouvatsos (1985). Further references related to the topics presented in this Chapter are given below.

Feed-forward configurations

Feed-forward configurations of the type analyzed in this Chapter were also considered by Lee and Pollock (1990). The queueing network analyzed is the same as the one considered in section 5.2. The algorithm yields the queue-length distribution for each node. It is a single-node decomposition algorithm and it consists of a forward pass and a backward pass. In the forward pass, the rate of arrivals into each node i is updated using information from its upstream nodes, which have already been visited. In the backward pass, the effective service time of each node i is updated using information from all its downstream nodes which have already been visited. The effective service time at each node is assumed to be exponentially distributed.

Takahashi (1989) analyzed a feed-forward queueing network with blocking-before-service assuming exponential services and Poisson arrivals. Each node was served by two servers. The configuration is similar to the one shown in figure 5.1. The rate matrix of this system has a special structure which is exploited by the approximation algorithm. Takahashi analyzed this queueing network using customer decomposition, two-node decomposition, and three-node decomposition.

Arbitrary configurations

Takahashi, Miyahara and Hasegawa (1980) and Labetoulle and Pujolle (1977) (1980) were probably the first to propose approximation algorithms for queueing networks with blocking where the nodes are linked in an arbitrary fashion. Takahashi, Miyahara and Hasegawa (1980) developed a single-node decomposition algorithm for the case where each node is served by a single exponential server and external arrivals occur to each node in a Poisson fashion. Blocking-after-service was assumed. The problem of deadlock that may arise in this case was not taken into account explicitly. For node i, let U_i be the set of its upstream nodes. Then, the effective arrival rate into node i is

$$\bar{\lambda}=\sum_{j\in U_i} r_{ji}\bar{\mu}_j)(1-p_j(0)),$$

where r_{ji} , $j\in U_i$, is the probability that a customer upon completion of its service at node j will join node i, $\bar{\mu}_j$ is the effective service time at the upstream node j, and $p_j(0)$ is the probability that node j is empty. The total arrival rate λ_i into node i is calculated using the expression $\lambda_i=\bar{\lambda}[1-p_i(m_i)]$, where $p_i(m_i)$ is the probability node i is full. The effective service time of the ith node $1/\bar{\mu}_i$ is obtained from the expression

$$\frac{1}{\bar{\mu}_i}=\frac{1}{\mu_i}+\sum_{k\in D_i} r_{ik}p_k(m_k)\Big[1+\sum_{\substack{j\in U_k\\ j\neq i}} \rho_j r_{jk}\frac{\bar{\mu}_j}{\bar{\mu}_j+\mu_i}\Big]\bar{\mu}_k,$$

where the second term reflects the blocking delay a customer may suffer upon completion of its service at the ith node. In particular, D_i is the set of downstream nodes of node i. For each of these nodes k, $k \in D_i$, the customer may be blocked for a period of time. The mean time of this period is given by the term in the square brackets. The summation term within the square brackets represents the additional delay the blocked customer suffers due to customers from other nodes which are also blocked and which will enter node k first. The quantity ρ_j used in this second summation term is the traffic intensity of the jth node. These equations can be written down for each node. The solution to the queueing network is obtained by solving the set of equations obtained from all nodes. The algorithm yields the blocking probability and its throughput for each node.

Labetoulle and Pujolle (1977) and (1980) analyzed an arbitrary configuration under repetitive-service blocking with fixed destination. The authors developed a single-node decomposition algorithm for exponentially distributed service times. They then extended the algorithm to the case of general distributions.

An approximation algorithm for analyzing an open queueing network of arbitrarily linked finite capacity queues with the deadlock detection and resolution scheme discussed in the introduction of section 5.3 was proposed by Jun and Perros (1989). The algorithm is an extension of the approximation procedure described in section 5.2. The service process at each node is revised to include a phase-type distribution, whose parameters reflect all the possible blocking delays and deadlocks. This phase-type distribution is rather complicated. After it is constructed, it is simplified by collapsing parts of it to C_2 distributions using the three moment approximation. In particular, each part of the phase-type distribution that represents the blocking delay due a node is collapsed into a C_2 distribution. Thus, the resulting effective mechanism has a simple structure and the node under study can be analyzed numerically. The approximation algorithm is an iterative scheme and it is similar in spirit to the isolation method of Labetoulle and Pujolle (1980). An iteration is completed when every node in the queueing network has been analyzed. Following a convergence test, the algorithm stops or another iteration is carried out. The order in which each node is analyzed in not important if the nodes are fully interconnected. Within each iteration, in order to analyze a node, one needs certain values from the other nodes. For those nodes that they have already been visited during the iteration, the new updated values are used. For the nodes that have not been visited yet, the values from the previous iteration are used.

The maximum entropy algorithm by Kouvatsos and Xenios (1989a) was also extended by the same authors to queueing networks with multiple server nodes under repetitive-service blocking with random destination or with fixed destination (see Kouvatsos and Xenios (1989b)). For extensions to other blocking mechanisms see Xenios (1989). A generalization of the maximum entropy algorithm to open queueing networks with repetitive-service blocking with random destination and multiple job classes was reported in Kouvatsos and Denazis (1993).

Kerbache and MacGregor Smith (1987) proposed an approximation method for analyzing queueing networks under repetitive-service blocking with fixed destination. Service times and external inter-arrival times were assumed to have a general distribution. The authors applied their method to small size tandem, split, and merge configurations. Akyildiz and von Brand (1989) obtained a product-form solution for open, closed, and mixed queueing networks assuming repetitive-service blocking with random destination and multiple classes. Upper and lower bounds on the throughput of open queueing networks with blocking were reported in Onvural and Perros (1988/89). Finally, Yamashita (1990) proposed an approximation algorithm for the analysis of an arbitrarily linked open queueing network, where all nodes share the same common memory.

Bibliography

Akyildiz, I. F. and von Brand, H., Exact solutions for open, closed and mixed queueing networks with rejection blocking, *Theoret. Comput. Sci.* **64** (1989) 203-219.

Altiok, T. and Perros, H. G., Approximate analysis of arbitrary configurations of open queueing networks with blocking, *Ann. Oper. Res.* **9** (1987) 481-509.

Jaynes, E. T., Information theory and statistical mechanics, *Phys. Rev.* **106** (1957) 620-630.

Jun, K.-P. and Perros, H. G., Approximate analysis of arbitrary configurations of queueing networks with blocking, in: Perros and Altiok, (Eds.), *Proc. First International Workshop on Queueing*

Networks with Blocking, (North-Holland, Amsterdam, 1989) 259-280.

Kerbache, L. and MacGregor Smith, J., The generalized expansion method for open finite queueing networks, *Eur. J. Oper. Res.* **32** (1987) 448-461.

Kouvatsos, D. D., Maximum entropy methods for general queueing networks, in: Potier, (Ed.), *Modelling techniques and tools for performance analysis*, (North-Holland, Amsterdam, 1985) 589-608.

Kouvatsos, D. D. and Denazis, S. G., *Comments on and tuning to: MEM for arbitrary queueing networks with repetitive-service blocking: A revision*, Tech. Rept. CS-18-91, Dept. of Computing, Univ. of Bradford, 1991.

Kouvatsos, D. D. and Denazis, S. G., Entropy maximized queueing networks with blocking and multiple job classes, *Perform. Eval.* **17** (1993) 189-205.

Kouvatsos, D. D. and Xenios, N. P., Maximum entropy analysis of general queueing networks with blocking, in: Perros and Altiok, (Eds.), *Proc. First International Workshop on Queueing Networks with Blocking*, (North-Holland, Amsterdam, 1989a) 281-309.

Kouvatsos, D. D. and Xenios, N. P., MEM for arbitrary queueing networks with multiple general servers and repetitive-service blocking, *Perform. Eval.* **10** (1989b) 169-195.

Labetoulle, J. and Pujolle, G., Modelling of packet switching communication networks with finite buffer size at each node, in: Chandy and Reiser, (Eds.), *Computer Performance*, (North Holland, Amsterdam, 1977) 515-535.

Labetoulle, J. and Pujolle, G., Isolation method in a network of queues, *IEEE Trans. Soft. Eng.* **6** (1980) 373-381.

Lee, H.-S. and Pollock, S. M., Approximate analysis of open acyclic exponential queueing networks with blocking, *Oper. Res.* **38** (1990) 1123-1134.

Onvural, R. O. and Perros, H. G., Equivalencies between open and closed queueing networks with finite buffers, *Perfom. Eval.* **9** (1988/89) 263-269.

Perros, H. G. and Snyder, P. M., A computationally efficient approximation algorithm for analyzing queueing networks with blocking, *Perform. Eval.* **9** (1988/89) 217-224.

Shannon, C. E., A mathematical theory of communications, *Bell System. Tech. J.* **27** (1948) 379-423, 622-656.

Takahashi, Y., Aggregate approximation for acyclic queueing networks with communication blocking, in: Perros and Altiok, (Eds.), *Proc. First International Workshop on Queueing Networks with Blocking*, (North-Holland, Amsterdam, 1989) 33-46.

Takahashi, Y., Miyahara, H. and Hasegawa, T., An approximation method for open restricted queueing networks, *Oper. Res.* **28** (1980) 594-602.

Xenios, N. P., *General Queueing Networks with Blocking*, Ph.D. Thesis, University of Bradford, 1989.

Yamashita, H., *Approximate aggregation for arbitrary configurations of queueing networks with shared buffer and blocking*, Tech. Rept., Mechanical Engineering Dept., Sophia University, 1990.

CHAPTER 6

Closed Queueing Networks With Blocking With Product-Form Solution

6.1. Introduction

In this Chapter, we examine closed queueing networks with blocking which have a product-form solution. In section 6.2, we summarize some relevant results on reversible Markov processes that we will make use of in the remaining of this Chapter. In section 6.3, we present results for two-node closed queueing networks with blocking with a single class of customers, and subsequently with multiple classes of customers. In section 6.4, we examine closed queueing networks with blocking with more than two nodes whose underlying Markov process is reversible. Two additional product-form results are given in section 6.5. Finally, relevant references are given in section 6.6.

6.2. Reversible Markov processes

In this section, we review some basic concepts on reversible Markov processes that we will use in this Chapter. Let X(t) be a stochastic process taking values in a countable space S for $t \in T$. Let $q(n,n')$ be the transition rate from state n to state n′. A stochastic process X(t) is reversible if $X(t_1)$, $X(t_2),\ldots,X(t_i)$ has the same distribution as $X(\tau-t_1)$, $X(\tau-t_2),\ldots,X(\tau-t_i)$ for all $t_1,t_2,\ldots,t_i$, $\tau \in T$. That is, if the direction of time is reversed the behaviour of the process remains the same. In view of this, a reversible Markov process is stationary. Let p(n) be the steady-state probability that the process is in state n, $n \in S$. Then, p(n), $n \in S$, satisfy the well known equations

$$p(n) \sum_{n' \in S} q(n,n') = \sum_{n' \in S} p(n')q(n',n), \; n \in S. \qquad (6.2.1)$$

Theorem 1: A stationary Markov process is reversible if and only if there exists a collection of positive numbers p(n), $n \in S$, summing to unit that satisfy the conditions

$$p(n)q(n,n') = p(n')q(n,n'), \ n,n' \in S. \qquad (6.2.2)$$

When there exists such a collection p(n), $n \in S$, it is the steady-state distribution of the process. That is, p(n), $n \in S$, also satisfy equations (6.2.1).

The quantity p(n)q(n,n′) is called the *probability flux*. Equations (6.2.2) are known as *detailed balance equations*. These equations require that the probability flux from state n to n′ equals the probability flux from state n′ to n. The full balance equations given by (6.2.1) require that the probability flux out of state n equals the probability flux into state n.

The reversibility of a stationary Markov process can be established from its steady-state distribution and its transition rates using equation (6.2.2). Alternatively, it could be established directly from its transition rates using the following theorem.

Theorem 2: A stationary Markov process is reversible if and only if its transition rates satisfy

$$q(n_1,n_2)q(n_2,n_3) \ldots q(n_{i-1},n_i)q(n_i,n_1)$$
$$= q(n_1,n_i)q(n_i,n_{i-1}) \ldots q(n_3,n_2)q(n_2,n_1)$$

for any finite sequence of states $n_1,n_2,\ldots,n_i \in S$.

An alternative way to establish reversibility of a Markov process is by checking the structure of its rate matrix. In particular, it can be shown that a stationary Markov process is reversible if and only if its rate matrix can be written as the product of a symmetric and a diagonal matrix.

Let us consider a reversible Markov process X(t) with state space S and steady-state distribution p(n), $n \in S$. If S is truncated to a sub-space A, then the resulting Markov process Y(t) is reversible in equilibrium and its steady-state distribution $p_A(n)$ is

$$p_A(n) = \frac{p(n)}{\sum_{a \in A} p(a)}, \ n \in A. \qquad (6.2.3)$$

We note that $p_A(n)$ is just the conditional probability that the original process is in state n given that it is somewhere in A. We say that a Markov process Y(t) is a *truncated process* of X(t) if (i) it is irreducible, and (ii) the state space of Y(t) is a subset of the state space of X(t) and the transition rates between the states of Y(t) are the same as the transition rates between the same states of X(t).

6.3. Two-node closed queueing networks with blocking

In this section, we examine a two-node closed queueing network with blocking under different types of blocking mechanisms. We first analyze this queueing network assuming a single class of customers, and subsequently we consider the case of multiple classes of customers.

Single class of customers

Let us consider the two-node closed queueing network with blocking as shown in figure 6.1. Let N be the number of customers in the network, and let m_1 and m_2 be the capacity of the first and second node respectively. We have, $N \leq m_1+m_2$. Service times at the first and second node are exponentially distributed with rate μ_1 and μ_2 respectively. Let $p(n_1,n_2)$ be the steady-state probability that the queueing network is in state (n_1,n_2), where n_1 and n_2 is the number of customers in the first and second node respectively. Since $n_1+n_2=N$, we have that $p(n_1,n_2)=p_1(n_1)$, where $p_1(n_1)$ is the marginal probability that there are n_1 customers in the first node. Below, we obtain $p_1(n_1)$ under blocking-before-service and blocking-after-service for the case where each queue is served by a single server. The results obtained here can be easily extended to the case where each queue is served by multiple servers.

a. *Blocking-before-service*

Under blocking-before-service, a server cannot start a service if the other node is full. We examine this model when $N<m_1+m_2$. We note that for $N=m_1+m_2$ we have a trivial solution since each node is always full and

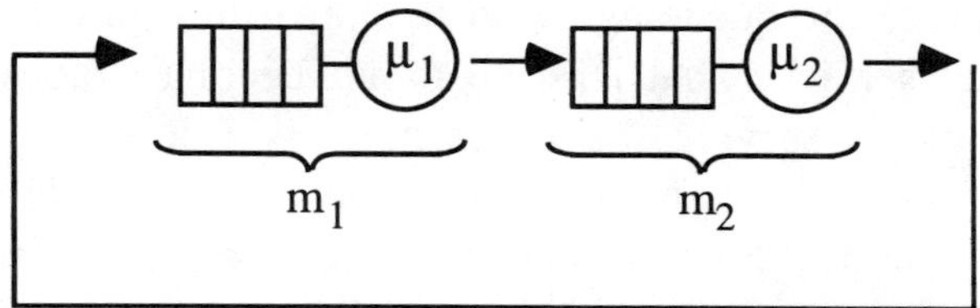

Figure 6.1: A two-node closed queueing network with blocking.

therefore each server is permanently blocked. Let $m_1,m_2 < N$, and define n_0 and n_u as the minimum respectively maximum occupancy in node 1. We have $n_0=\max\{0,N-m_2\}$, $n_u=\min\{N,m_1\}$ and $n_0 \le n_1 \le n_u$. The rate diagram associated with node 1 has the following simple structure

$$(n_0) \underset{\mu_1}{\overset{\mu_2}{\rightleftarrows}} (n_0+1) \underset{\mu_1}{\overset{\mu_2}{\rightleftarrows}} (n_0+2) \underset{\mu_1}{\overset{\mu_2}{\rightleftarrows}} \cdots \underset{\mu_1}{\overset{\mu_2}{\rightleftarrows}} (n_u)$$

Thus, we can easily obtain that

$$p_1(n_1) = \left(\frac{\mu_2}{\mu_1}\right)^{n_1-n_0} p_1(n_0),\ n_0 \le n_1 \le n_u, \tag{6.3.1}$$

where

$$p_1(n_0) = \begin{cases} \dfrac{1-(\mu_2/\mu_1)}{1-(\mu_2/\mu_1)^{n_u-n_0+1}} & \text{if } \mu_2 \ne \mu_1 \\[2ex] \dfrac{1}{n_u-n_0+1} & \text{if } \mu_2=\mu_1 . \end{cases} \tag{6.3.2}$$

If $m_1,m_2 \ge N$, then $n_0=0$ and $n_u=N$. In this case, the queue-length distribution of the first node becomes identical to that of an M/M/1/N queue with an arrival and service rate equal to μ_2 and μ_1 respectively. If $m_1 \le N \le m_2$, then $n_0=0$ and $n_u=m_1$. In this case, the queue-length distribution of the first node becomes identical to that of an M/M/1/m_1 queue with an arrival and service rate equal to μ_2 and μ_1 respectively. Finally, we note that (6.3.1) and (6.3.2) also hold under the repetitive-service blocking mechanism.

b. Blocking-after-service

We now analyze the queueing network shown in figure 6.1 assuming blocking-after-service. That is, a customer becomes blocked if upon completion of its service the other node is full. We assume that $N<m_1+m_2$. As before, for $N=m_1+m_2$ we have a trivial solution since in this case both nodes are full, and therefore, each server will become blocked immediately after it completes its first service. Let n_0 and n_u be the minimum and maximum occupancy of node 1, ie. $n_0=\max(0,N-m_2)$

and $n_u=\min\{N,m_1\}$. If $m_1,m_2<N$, we have that $n_0>0$, $N-n_0=m_2$, and $n_u=m_1$. Then, the rate diagram associated with this system is

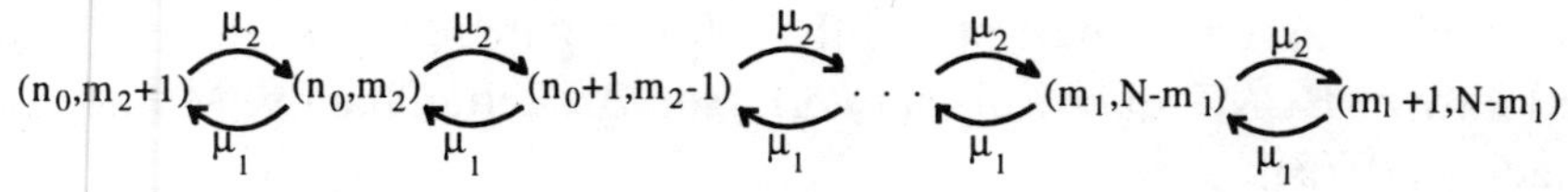

$$(n_0,m_2+1) \underset{\mu_1}{\overset{\mu_2}{\rightleftarrows}} (n_0,m_2) \underset{\mu_1}{\overset{\mu_2}{\rightleftarrows}} (n_0+1,m_2-1) \underset{\mu_1}{\overset{\mu_2}{\rightleftarrows}} \cdots \underset{\mu_1}{\overset{\mu_2}{\rightleftarrows}} (m_1,N-m_1) \underset{\mu_1}{\overset{\mu_2}{\rightleftarrows}} (m_1+1,N-m_1)$$

The states (n_0,m_2+1) and $(m_1+1,N-m_1)$ are associated with the first server respectively the second server being blocked. We can easily obtain that

$$p(n_0+i,m_2-i) = \left(\frac{\mu_2}{\mu_1}\right)^{i+1} p(n_0,m_2+1),\ i=0,1,\ldots,m_1-n_0, \qquad (6.3.3)$$

$$p(m_1+1,N-m_1) = \left(\frac{\mu_2}{\mu_1}\right)^{m_1-n_0+1} p(n_0,m_2+1) \qquad (6.3.4)$$

and

$$p_1(n_0,m_2+1) = \begin{cases} \dfrac{1-(\mu_2/\mu_1)}{1-(\mu_2/\mu_1)^{m_1-n_0+3}} & \text{if } \mu_2 \neq \mu_1 \\[2ex] \dfrac{1}{m_1-n_0+3} & \text{if } \mu_2=\mu_1. \end{cases} \qquad (6.3.5)$$

If $m_1 \le N \le m_2$, then the first server can never get blocked seeing that the capacity of the second node is large enough to accommodate all N customers. The rate diagram of the system can be obtained from the one given above by ignoring the blocking state (n_0,m_2+1) and setting $n_0=0$ and $n_u=m_1$. The solution has a similar form as (6.3.3) to (6.3.5). We note that the first node becomes identical to an $M/M/1/m_1+1$ queue with arrival and service rate equal to μ_2 and μ_1 respectively. If $m_1,m_2 \ge N$, then neither server can get blocked. Therefore, the first node becomes identical to an M/M/1/N queue with an arrival and service rate equal to μ_2 and μ_1 respectively.

Multiple classes of customers

Let us now consider a two-node queueing network with blocking and C classes of customers. As before, let m_i be the capacity of the ith node, i=1,2, and N the total number of customers in the queueing network. We assume that $m_i<N$ and that $m_1+m_2>N$. A class c customer in node i may change to class d when it joins node j with probability $r_{ic,jd}$, where

$$\sum_{jd} r_{ic,jd} = 1, \; c=1,2,\ldots,C.$$

Each node is assumed to be one of the following three types allowed in the well known BCMP queueing network (see Baskett, Chandy, Muntz, and Palacios (1975)):

Type 1: There is a single server whose speed may depend upon the total number of customers in the node. All classes receive an exponentially distributed service with the same mean. Customers are served in FIFO order.

Type 2: There is a single server whose speed may depend on the total number of customers in the node. Each class receives a service which has a Coxian distribution with class dependent parameters. All customers in the node are processor-shared.

Type 3: This is an infinite capacity server queue. That is, for each customer in the node there is a server available to serve the customer. Each class receives a service which has a Coxian distribution with class dependent parameters.

The state of this two-node queueing network can be described by $(\mathbf{n}_1,\mathbf{n}_2)$, where vector $\mathbf{n}_i$ represents the state of the ith node, i=1,2. Let Y(t) be the underlying Markov process of this queueing network and let A be its state space. Using the notion of truncated reversible Markov processes we show below that this two-node queueing network under blocking-before-service, or blocking-after-service, or repetitive-service blocking, has a product-form solution.

Let us consider the two-node queueing network under study assuming that each node has an infinite capacity queue. We note that this queueing network is of the BCMP type. Also, its underlying Markov process, X(t), is reversible. (This can be verified by showing that the detailed balance equations (6.2.2) hold.) Let $(\mathbf{n}_1',\mathbf{n}_2')$ be its state, where $\mathbf{n}_i'$ is the state of node i, i=1,2, and let S be its state space. Below, we show that Y(t) is a truncated process of X(t).

Let us first consider Y(t) assuming blocking-after-service. Now, truncate S into a sub-space B such that at most m_i+1 customers are allowed in node i, i=1,2. Then, as it was mentioned in section 6.2., the resulting Markov process is also reversible. This truncated Markov process is in fact Y(t). This can be easily shown by first establishing that the state spaces A and B are identical. Let $\mathbf{n}_i^b$ be a state of Y(t) denoting that the ith node is blocked by the other node. Define the following mapping from the states of Y(t) to the states of the truncated process

$$(\mathbf{n}_1,\mathbf{n}_2) = (\mathbf{n}_1',\mathbf{n}_2'), \text{ if no node is blocked}$$

$$(\mathbf{n}_1^b,\mathbf{n}_2) = (\mathbf{n}_1'\text{-b},\mathbf{n}_2'\text{+b}), \text{ if node 1 is blocked by node 2}$$

$$(\mathbf{n}_1,\mathbf{n}_2^b) = (\mathbf{n}_1'\text{+b},\mathbf{n}_2'\text{-b}), \text{ if node 1 is blocked by node 2,}$$

where $\mathbf{n}_i'$-b is the same as $\mathbf{n}_i^b$ with the blocked customer removed and $\mathbf{n}_i'$+b is the same as $\mathbf{n}_i$ with the blocked customer added to the queue. We note that the above mapping is one to one and onto. Hence, the state spaces B and A are identical. Also, it can be easily verified that the transition rates between the equivalent states of the two processes are the same. Thus, the above two-node queueing network with blocking has a product-form equilibrium distribution, which can be obtained using expression (6.2.3.)

Let us consider Y(t) assuming blocking-before-service. Truncate S into a sub-space B such that at most m_i customers are allowed in node i, i=1,2. The resulting truncated process is reversible and it is identical to Y(t). This can be shown by following the same arguments as above. State spaces B and A are identical, since each state $(\mathbf{n}_1',\mathbf{n}_2')$ of B is identical to a state $(\mathbf{n}_1,\mathbf{n}_2)$ of A. The transition rates between identical states can be shown to be the same. The same argument holds for repetitive-service blocking.

6.4. Closed queueing networks with blocking with more than two nodes

In this section, we examine closed queueing networks with blocking with more than two nodes whose underlying Markov process is reversible. We first consider two single class queueing networks, namely, a queueing networks with reversible routing matrix, and a queueing network with state-dependent routing. Subsequently, we examine an open, closed, and mixed queueing network with multiple classes and reversible routing.

Reversible routing

Let us consider a queueing network consisting of M nodes arbitrarily linked. Let N be the total number of customers in the system. A customer at the ith node, i=1,2,...,M, requires an exponentially distributed service with mean $1/\mu_i$. Let $f_i(n_i)$ be the service capacity of the server at the ith node, when there are n_i customers in the node. We have, $f_i(n_i)>0$ if $n_i>0$, and $f_i(n_i)=0$ if $n_i=0$. The probability that a customer in node i will attempt to join node j upon completion of its service is r_{ij}. Let $b_j(n_j)$ be the probability that a customer will be admitted to node j when there are n_j customers in the node. If m_j is the maximum capacity of node j, then $b_j(m_j)=0$. The blocking mechanism is repetitive service with random destination. That is, if a customer at node i finds that its selected destination node is full at the moment it completes its service, then the customer receives another service at node i. It then attempts to enter a new destination node chosen independently of the destination selected the previous time. This is repeated until the customer successfully enters a destination node. Below, we show that this queueing network has a product-form solution under the assumption that the routing matrix (r_{ij}) is reversible.

Let $\mathbf{n}=(n_1,n_2,\ldots,n_M)$ be the state of the queueing network, where n_i is the number of customers in the ith node, and let $p(\mathbf{n})$ be the steady-state probability of being in state $\mathbf{n}$. Then, we have the steady-state equations

$$p(\mathbf{n}) \sum_{i=1}^{M} \sum_{j=1}^{M} \mu_i f_i(n_i) b_j(n_j) r_{ij} = \sum_{i=1}^{M} \sum_{j=1}^{M} p(\mathbf{n}+\mathbf{1}_i-\mathbf{1}_j) \mu_i f_i(n_i+1) b_j(n_j-1) r_{ij}$$

for $\mathbf{n}\in S$, where S is the set of all feasible states. The compact notation $\mathbf{n}+\mathbf{1}_i-\mathbf{1}_j$ is used to indicate the vector $(n_1,\ldots,n_i+1,\ldots,n_j-1,\ldots,n_M)$. That is, $\mathbf{1}_i$ is a row vector of zeroes except the ith column which is 1. For simplicity, we assume that the summation over indices i and j is such that $i\neq j$.

If the routing matrix (r_{ij}) is reversible, then there exist positive e_i, $i=1,2,\ldots,M$, which satisfy the equation

$$e_i r_{ij} = e_j r_{ji} \tag{6.4.1}$$

for all i and j. In this case, it can be shown that

$$p(\mathbf{n}) = \frac{1}{G} \prod_{i=1}^{M} \prod_{k=1}^{n_i} \frac{e_i\, b_i(k-1)}{\mu_i\, f_i(k)}, \quad \mathbf{n}\in S, \tag{6.4.2}$$

where G is the normalizing constant. This can be easily demonstrated by considering the detailed balance equations

$$p(\mathbf{n})\mu_i f_i(n_i) b_j(n_j) r_{ij} = p(\mathbf{n}-\mathbf{1}_i+\mathbf{1}_j)\mu_j f_j(n_j+1) b_i(n_i-1) r_{ji}, \quad \mathbf{n}\in S.$$

Using (6.4.1) we have

$$p(\mathbf{n})\mu_i f_i(n_i) b_j(n_j) = p(\mathbf{n}-\mathbf{1}_i+\mathbf{1}_j)\mu_j f_j(n_j+1) b_i(n_i-1)\frac{e_j}{e_i}, \ \mathbf{n}\in S. \quad (6.4.3)$$

One can verify that expression (6.4.2) for p($\mathbf{n}$) satisfies all the detailed balance equations. Therefore, if the routing matrix is reversible, the underlying stationary Markov process of the queueing network is reversible and the probability p($\mathbf{n}$) that the queueing network is in state $\mathbf{n}$ has the product-form solution given by (6.4.2).

For a particular queueing network, we can check whether its routing matrix is reversible by applying theorem 2 (adopted for Markov chains). Alternatively, we can check whether the matrix can be written out as a product of a symmetric and a diagonal matrix. Using the latter approach, the following type of routing is reversible

$$r_{ij} = \psi_j/\phi_i, \ i\neq j, \ r_{ii}=0, \quad (6.4.4)$$

where ψ_j and ϕ_i are all positive. This can be easily shown by setting the (i,j)th element of the symmetric matrix equal to $(\phi_i\phi_j)^{-1}$, $i\neq j$, the diagonal elements of the symmetric matrix equal to zero, and the elements of the diagonal matrix equal to $\psi_i\phi_i$.

State-dependent routing

We now consider the case where the routing probability from node i to node j is state dependent. In particular, let us consider the queueing network studied above assuming that r_{ij} depends on n_i and n_j through the expression

$$r_{ij}(n_i,n_j) = \phi_i(n_i)\psi_j(n_j), \quad (6.4.5)$$

where $\phi_i(.)$ and $\psi_j(.)$ are arbitrary functions, such that $\phi_i(n_i)>0$ if $n_i>0$, $\psi_j(n_j)>0$ if $0\leq n_j<m_j$, $\phi_i(0)=0$ and $\psi_j(m_j)=0$. For simplicity, $i\neq j$. In this case the steady-state equations become

$$p(\mathbf{n})\sum_{i=1}^{M}\sum_{j=1}^{M}\mu_i f_i(n_i)\phi_i(n_i)\psi_j(n_j)$$

$$= \sum_{i=1}^{M}\sum_{j=1}^{M} p(\mathbf{n}-\mathbf{1}_i+\mathbf{1}_j)\mu_i f_j(n_j+1)\phi_j(n_j+1)\psi_i(n_i-1)$$

for $\mathbf{n} \in S$, where S is the set of all feasible states. Under the above state-dependent routing probability the underlying Markov process of the queueing network is reversible and $p(\mathbf{n})$ is

$$p(\mathbf{n}) = \frac{1}{G} \prod_{i=1}^{M} \prod_{k=1}^{n_i} \frac{\psi_i(k-1)}{\mu_i f_i(k) \phi_i(k)}, \tag{6.4.6}$$

where G is the normalizing constant. This can be easily demonstrated by verifying that the above solution satisfies the detailed balance equations

$$p(\mathbf{n})\mu_i f_i(n_i)\phi_i(n_i)\psi_j(n_j) = p(\mathbf{n}-\mathbf{1}_i+\mathbf{1}_j)\mu_j f_j(n_j+1)\phi_j(n_j+1)\psi_i(n_i-1),\ \mathbf{n} \in S,\ i \neq j.$$

This state-dependent routing has the following interpretation. A customer, upon completion of its service at node i, will probabilistically join any of the destination nodes which at that moment are not full. If all destination nodes are full, then the customer will repeat its service at node i (as in the repetitive service with random destination blocking mechanism).

We note that the routing probability $r_{ij}(n_i,n_j)$ should satisfy $\sum_j r_{ij}(n_i,n_j)=1$, or $\sum_j \phi_i(n_i)\psi_j(n_j)=1$. From this we obtain

$$\phi_i(n_i) = \left[\sum_{\substack{j=1 \\ j \neq i}}^{M} \psi_j(n_j) \right]^{-1}.$$

Now, let us set $\psi_j(n_j)=m_j-n_j$. Then

$$\phi_i(n_i) = \left[\left(\sum_{\substack{j=1 \\ j \neq i}}^{M} m_j \right) - (N-n_i) \right]^{-1}.$$

Thus, we have the following *probabilistic shortest queue* routing

$$r_{ij}(n_i,n_j) = \frac{m_j - n_j}{\left(\sum_{\substack{j=1 \\ j \neq i}}^{M} m_j \right) - (N-n_j)}, \quad i \neq j . \tag{6.4.7}$$

In this type of routing a job may join a node which has the shortest queue with the highest probability. A job never joins a node which is full.

When the routing matrix is not reversible, or when the routing probabilities are not state-dependent, the above queueing network can be shown to have a product-form solution in the following two cases:

a. The probability $b_j(n_j)$ that a customer will be admitted to node j is constant, independent of n_j. That is, $b_j(n_j)=b_j$, j=1,2,...,M.
b. The rate $f_i(n_i)$ at which server i works is constant, independent of n_i. That is, $f_i(n_i)=f_i$, i=1,2,...,M. Since $f_i(0)=0$ and $f_i(n_i)>0$ for $n_i>0$, it is necessary to assume that the case $n_i=0$ never arises. That is, none of the nodes can be empty. We note that for cyclic queueing networks the solution is identical to the solution given below in section 6.5 for a cyclic queueing network for which $N \geq \sum_{i=1}^{M} m_i - \min_i\{m_i\}$.

Multiple classes of customers

So far, the results presented in this section have been obtained assuming a single class of customers. We now consider a queueing network with multiple classes, reversible routing, and repetitive service with random destination blocking, whose underlying Markov process is reversible.

We assume that the queueing network under study consists of M nodes, arbitrarily linked, and C classes. Let us first consider a single node, say node i, in isolation. A class c customer requests a service at node i which has a distribution F_{ic} with mean $1/\mu_{ic}$. This distribution is assumed to be a finite mixture of Erlang distributions, that is

$$F_{ic} = \sum_{p} g_{ic,p} E_{p\nu_{ic}}, \tag{6.4.8}$$

where $E_{p\nu_{ic}}$ denotes an Erlang distribution with p phases, each with a node and class dependent rate ν_{ic}. The above definition basically means that a class c customer requests an Erlang service time $E_{p\nu_{ic}}$ with probability $g_{ic,p}$. Obviously, $\Sigma_p g_{ic,p}=1$. We also have that

$$\frac{1}{\mu_{ic}} = \sum_{p} g_{ic,p} \frac{1}{\nu_{ic}}.$$

From renewal theory, we have that the probability that at an arbitrary instant a class c customer still has to traverse p′ phases is

$$R_{ic}(p') = \frac{\mu_{ic}}{\nu_{ic}} \sum_{p \geq p'} g_{ic,p} .$$

Note that $R_{ic}(1)=\mu_{ic}/\nu_{ic}$.

The state of node i will be denoted by the vector $\mathbf{n}_i=((c_{i1},\sigma_{i1}),(c_{i2},\sigma_{i2}),\ldots,(c_{in_i},\sigma_{in_i}))$, where c_{il} is the class of the customer in node i occupying position l, σ_{il} is the number of remaining phases of service of that customer, and n_i is the total number of customers in node i. The total number of class c customers in node i will be denoted by n_{ic}. The scheduling discipline at node i is described by the tuplet (f_i,γ_i,ξ_i).

$f_i(n_i)$ is the service capacity of the server when there are n_i customers in the node.

$\gamma_i(l,n_i)$ is the fraction of the service capacity destined to the customer in position l, $l=1,2,\ldots,n_i+1$, when there are n_i customers in node i, $\sum_l \gamma_i(l,n_i)=1$. When the customer in the lth position departs from the node, the customers in positions $l+1,l+2,\ldots,n_i$ move to positions $l,l+1,l+2,\ldots,n_i-1$ respectively.

$\xi_i(l,n_i)$ is the probability that an arriving customer to node i is placed in position l, $l=1,2,\ldots,n_i+1$, when there are n_i customers, $\sum_l \xi_i(l,n_i)=1$. The customers previously occupying positions $l,l+1,\ldots,n_i$, move to positions $l+1,l+2,\ldots,n_i+1$ respectively.

This scheduling discipline is quite general, and it can be used to describe some of the well known disciplines. For instance, a node with k identical servers and first-in-first-out can be described by setting $f_i(n_i)=\min(n_i,k)$, $\gamma_i(1,n_i)=1$ and $\xi_i(n_i+1,n_i)=1$. A single server node with last-in-first-out can be described by setting $f_i(n_i)=1$, $\gamma_i(n_i,n_i)=1$ and $\xi_i(n_i+1,n_i)=1$, and with processor-sharing by setting $\gamma_i(l,n_i)=1/n_i$ and $\xi_i(n_i+1,n_i)=1$. The scheduling discipline (f_i,γ_i,ξ_i), however, cannot describe disciplines which involve priorities, such as head-of-line. The scheduling discipline is called *symmetric* if $\gamma_i(l,n_i)=\xi_i(l,n_i+1)$.

For the queueing network considered here we assume a symmetric scheduling discipline, if the service time distribution is either class dependent or non-exponential. In particular, each node is assumed to satisfy one of the following: (i) The service time is exponentially distributed with a mean that does not depend on the customer's class. The scheduling discipline is arbitrary within the class of disciplines described above; (ii) The service distribution is given by (6.4.8) and it may be class dependent. A symmetric scheduling discipline is assumed.

Let us consider now the entire queueing network. The state of the queueing network will be described by the vector $\mathbf{n}=(\mathbf{n}_1,\mathbf{n}_2,\ldots,\mathbf{n}_M)$, where $\mathbf{n}_i$ is the state of node i described above. Let $r_{ic,jd}$ be the probability that a class c customer departing from node i joins node j as a class d customer. Note that $r_{ic,id}=0$ for all i, c and d. All external customers arrive in a single Poisson stream at the rate λ. Upon arrival, an external customer attempts to enter node i as a class d customer with probability $r_{0,jd}$. Let $r_{ic,0}$ be the probability that a class c customer that completed its service at node i leaves the network.

Following the BCMP convention, we consider the pair (node, customer class) known as job state. A customer is typically associated with several such pairs as it traverses a queueing network. The set of all pairs (node, class) is partitioned into one or more disjointed sets as follows. Two pairs (i,c) and (j,d) belong to the same set if a customer that can visit node i as class c can also visit node j as class d. Each of these sets is called a *chain.* A customer within a chain Γ may change to any job state that belongs to the same chain Γ when it moves through the queueing network. However, it may not change to a job state that belongs to another chain. If customers never change class when they go from node to node, then there will be as many chains as classes. A chain may be open or closed. In an open chain customers arrive from the outside and eventually depart from the queueing network. That is, for at least one node i and one class c in the chain we have that $r_{0,ic}\neq 0$, and also for at least one node j and a class d we have $r_{0,jd}\neq 0$. In a closed chain no departures to the outside or arrivals from outside are permitted. The queueing network considered here is assumed to be *mixed*, that is, it consists of both open and closed chains.

For each chain Γ, define e_{ic} as the solution to the system of equations

$$e_{ic} = \lambda r_{0,ic} + \sum_{(j,d)\in\Gamma} e_{jd} r_{jd,ic}, \quad (i,c)\in\Gamma.$$

For an open chain, the above system of equations has a unique solution, and the quantity e_{ic} is the throughput of class c customers at node i. For a closed chain, the above linear system is homogeneous. In this case, we arbitrarily fix one of the unknown e_{ic} and then solve for the remaining unknowns. In view of this, the quantities e_{ic} are known as *relative throughputs*.

We assume that the routing matrix $(r_{ic,jd})$ of each chain Γ is reversible. That is,

$$e_{ic}r_{ic,jd} = e_{jd}r_{jd,ic}, \qquad \text{for all } (i,c)(j,d) \in \Gamma$$

$$\lambda r_{0,jd} = e_{jd}r_{jd,0}, \qquad \text{for all } (j,d) \in \Gamma.$$

The probability that an arriving class c customer is allowed to enter node i is assumed to be a function of the number of class c customers n_{ic} present at the node, the number of customers $n_{i\Gamma}$ who belong to the same chain Γ, and the total number of customers n_i in the node. That is,

$$b_{ic}(n_{ic},n_{i\Gamma},n_i) = h_{ic}(n_{ic})h_{i\Gamma}(n_{i\Gamma})h_i(n_i),$$

where h_{ic}, $h_{i\Gamma}$ and h_i are arbitrary functions, such that if for a given value of n_i, say n_i', $h_i(n_i')=0$ then $h_i(n_i)=0$ for all $n_i \geq n_i'$. A similar constraint applies to h_{ic} and $h_{i\Gamma}$. When a customer is blocked from entering a destination node, the customer returns back to node i. The customer is treated like a new arrival, only it cannot be rejected, and it is placed in a position in the node according to the scheduling discipline.

We now consider the transitions out of a state **n** due to a specific customer. We assume that the customer occupies the *l*th position in the ith node, it is of class c, it has p phases of service left, and it belongs to chain Γ, i.e. $(i,c) \in \Gamma$. We have the following transitions associated with this customer.

The rate at which the customer advances to the next phase (assuming that p>1) is

$$\nu_{ic}f_i(n_i)\gamma_i(l,n_i).$$

The rate at which the customer completes its last phase (assuming that p=1), successfully enters node j as a class d customer, $(j,d) \in \Gamma$, occupies the l' position, and requests an Erlang service distribution with p' phases is

$$\nu_{ic}f_i(n_i)\gamma_i(l,n_i)r_{ic,jd}b_{jd}(n_{jd},n_{j\Gamma},n_j)\xi_j(l',n_j)g_{jd,p'}\,.$$

The rate at which the customer completes its last phase (assuming that p=1), fails to enter node j as class d customer, $(j,d) \in \Gamma$, returns to the l'

position of ith node and requests an Erlang service distribution with p′ phases is

$$\nu_{ic} f_i(n_i) \gamma_i(l,n_i) \sum_{(j,d)\in\Gamma} [r_{ic,jd}(1-b_{jd}(n_{jd},n_{j\Gamma},n_j))\xi_j(l',n_i-1)g_{ic,p'}] \,.$$

The rate at which the customer departs from the queueing network (assuming that p=1) is

$$\nu_{ic} r_{ic,0} f_i(n_i) \gamma_i(l,n_i).$$

The rate at which an external arrival successfully enters the ith node as class c customer, occupies the lth position, and requests an Erlang service distribution with p phases is

$$\lambda r_{0,ic} b_{ic}(n_{ic},n_{i\Gamma},n_i)\xi_i(l,n_i)g_{ic,p} \,.$$

Let S be the set of all feasible states of the queueing network, and let p(**n**) be the steady-state probability of being in state **n**, **n**∈ S. Using the above transition rates we can write down the steady state equations given by (6.2.1). It can be shown that the underlying Markov process of this queueing network is reversible, and that the probability p(**n**) that the queueing network is in state **n**∈ S is

$$p(\mathbf{n}) = \frac{1}{G}\prod_{i=1}^{M}[\prod_{l=1}^{n_i} \frac{h_i(l-1)}{f_i(l)} R_{ic_{il}}(\sigma_{il}) \prod_{\Gamma}\prod_{l=1}^{n_{i\Gamma}} h_{i\Gamma}(l-1) \prod_{c=1}^{C}\prod_{l=1}^{n_{ic}} \frac{e_{ic}h_{ic}(l-1)}{\mu_{ic}}], \qquad (6.4.9)$$

where Γ ranges over all routing chains, and G is the normalizing constant. We note that this solution holds for any arbitrary distribution with the same mean (see Barbour (1976)). That is, the solution is insensitive to the distribution of the service times and it only depends on the mean of the service times.

The vector **n** gives a detailed description of the class of the customer and its remaining phases of services that occupies the lth position at node i, l=1,2,...,n_i, i=1,2,...,M. This state can be aggregated so that only the class of the customer is given. Define $\mathbf{v}=(\mathbf{v}_1,\mathbf{v}_2,...,\mathbf{v}_M)$, where $\mathbf{v}_i=(c_{i1},c_{i2},...,c_{in_i})$ is a vector that gives the class of the customer that occupies the lth position in the ith node, l=1,2,...,n_i. Let p(**v**) be the steady-state probability that the queueing network is in state **v**. We have

$$p(\mathbf{v}) = \frac{1}{G} \prod_{i=1}^{M} p_i(\mathbf{v}_i), \tag{6.4.10}$$

where

$$p_i(\mathbf{v}_i) = \prod_{l=1}^{n_i} \frac{h_i(l-1)}{f_i(l)} \prod_{r} \prod_{l=1}^{n_{ir}} h_{ir}(l-1) \prod_{c=1}^{C} \prod_{l=1}^{n_{ic}} \frac{e_{ic} h_{ic}(l-1)}{\mu_{ic}}. \tag{6.4.11}$$

Expression (6.4.10) can be easily obtained by summing p(**n**) over all possible number of phases left for each customer, and observing that

$$\sum_p R_{ic}(p) = \sum_p \frac{\mu_{ic}}{v_{ic}} \sum_{k \geq p} g_{ic,k}$$

$$= \mu_{ic} \sum_p \frac{p g_{ic,p}}{v_{ic}}$$

$$= \mu_{ic} \frac{1}{\mu_{ic}}$$

$$= 1.$$

State **v** can be further aggregated to a state **w** where, for each node, only the number of customers of each class present in the node is given. Define $\mathbf{w}=(\mathbf{w}_1,\mathbf{w}_2,\ldots,\mathbf{w}_M)$, where $\mathbf{w}_i=(n_{i1},n_{i2},\ldots,n_{iC})$. For a given state $\mathbf{w}_i$ there are

$$\binom{n_i}{n_{i1}\ n_{i2}\ \ldots\ n_{iC}}$$

different ways of ordering these customers, where n_i is the total number of customers in the ith node. Since the solution for p(**v**) does not depend on the order of the customers in the nodes, we have

$$p(\mathbf{w}) = \frac{1}{G} \prod_{i=1}^{M} \binom{n_i}{n_{i1}\ n_{i2}\ \ldots\ n_{iC}} p_i(\mathbf{v}_i). \tag{6.4.12}$$

Calculation of the normalizing constant G

The calculation of the steady-state probability p(**n**) given by (6.4.2) requires knowledge of the normalizing constant G. Since $\sum_{\mathbf{n}\in S} p(\mathbf{n})=1$, we have that

$$G = \sum_{\mathbf{n}\in S} \left[\prod_{i=1}^{M} \prod_{k=1}^{n_i} \frac{e_i\, b_i(k-1)}{\mu_i\, f_i(k)} \right]. \tag{6.4.13}$$

In order to calculate G from (6.4.13) we have to enumerate all the feasible states, then calculate the term within the squared brackets for each state, and finally sum up all the terms. This brute-force approach works well when the number of states is very small. In general, however, one needs an efficient and stable method for calculating G. Such methods have been proposed in the literature for BCMP type of networks. Below, we describe a method for calculating G that is based on the well-known convolution algorithm for BCMP queueing networks.

Let us consider the single class queueing network analyzed at the beginning of section 6.4. We assume that the queueing network consists of M nodes arbitrarily linked and that the total number of customers in the network is N. For each node i we assume a capacity m_i, exponentially distributed service times with mean $1/\mu_i$, and FIFO service discipline. The probability that a customer in node i will attempt to join node j upon completion of its service is r_{ij}. The routing matrix (r_{ij}) is assumed to be reversible. That is, there exist positive e_i, i=1,2,...,M, which satisfy the equation $e_i r_{ij}=e_j r_{ji}$. Repetitive-service with random destination blocking is assumed. Also, we assume that $f_i(n_i)=1$ for $n_i \geq 0$, $b_j(n_j)=1$ if $n_i<m_i$ and $b_i(m_i)=0$. The state of this queueing network is given by the vector $\mathbf{n}=(n_1,n_2,\ldots,n_M)$, where n_i is the number of customers in the ith node, $n_i=0,1,\ldots,m_i$, and $\sum n_i=N$. The steady-state probability p(**n**), **n**$\in$ S, given by (6.4.2) has the following simple expression

$$p(\mathbf{n}) = \frac{1}{G} \prod_{i=1}^{M} \rho_i^{n_i}, \tag{6.4.14}$$

where $\rho_i=\lambda_i/\mu_i$.

Let us assume for a moment that the capacity of each node is infinite. Then, the state of the queueing network is given by the same vector $\mathbf{n}=(n_1,n_2,\ldots,n_M)$, where n_i takes the values $n_i=0,1,\ldots,N$ and $\sum n_i=N$, and the steady-state probability p(**n**) is given by (6.4.14). For this queueing network, the normalizing constant G can be calculated as follows (see Gelenbe and Mitrani (1980)). We first note that G is the sum of all terms $\rho_1^{n_1}\rho_2^{n_2}\ldots\rho_M^{n_M}$ for which $n_i=0,1,\ldots,N$ and $\sum n_i=N$. Consider now the generating functions

$$g_i(z)=\sum_{n_i=0}^{\infty}\rho_i^{n_i}z^{n_i},\ i=1,2,\ldots,M,$$

and the product

$$g(z)=\prod_{i=1}^{M}g_i(z)=\prod_{i=1}^{M}\Big(\sum_{n_i=0}^{\infty}\rho_i^{n_i}z^{n_i}\Big). \tag{6.4.15}$$

Clearly, G is the coefficient of z^N in g(z) since this coefficient is also the sum of all terms $\rho_1^{n_1}\rho_2^{n_2}\ldots\rho_M^{n_M}$ for which $n_i=0,1,\ldots,N$ and $\sum n_i=N$. This coefficient can be obtained recursively as follows. Define by $\tilde{g}_i(z)$ the partial products in (6.4.15), i. e. $\tilde{g}_i(z)=g_1(z)g_2(z)\ldots g_i(z)$. We have $\tilde{g}_1(z)=g_1(z)$ and

$$\tilde{g}_i(z)=\tilde{g}_{i-1}(z)g_i(z),\ i=2,3,\ldots,M. \tag{6.4.16}$$

Let G(i,j) be the coefficient of z^j in $\tilde{g}_i(z)$. Then, G=G(M,N). Since,

$$g_i(z)=\sum_{n_i=0}^{\infty}\rho_i^{n_i}z^{n_i}=1/(1-\rho_i z),$$

we can re-write (6.4.16) as

$$\tilde{g}_i(z)=\tilde{g}_{i-1}(z)+\rho_i z g_i(z),\ i=2,3,\ldots,M.$$

This implies the following relation for the coefficients G(i,j)

$$G(i,j)=G(i,j-1)+\rho_j G(i-1,j),\ i\geq 2,\ j\geq 1, \tag{6.4.17}$$

with initial values $G(0,j)=1$, $j\geq 1$, and $G(i,1)=\rho_1^i$, $i\geq 1$.

In the above calculation of G, terms of $\rho_1^{n_1}\rho_2^{n_2}\ldots\rho_M^{n_M}$ where $n_i>m_i$ were also included as long as $\sum n_i=N$. However, these terms are infeasible when the capacity of the nodes is finite. The following modified recursive procedure can be used to subtract the infeasible terms.

Use (6.4.17) for $j=2,3,\ldots,M$ and $i=1,2,\ldots,m_j$ with the initial conditions $G(0,j)=1$, $j=1,2,\ldots,M$, and

$$G(i,1) = \begin{cases} \rho_1^i & \text{if } i \le m_1 \\ 0 & \text{if } i > m_1 \end{cases} \quad , \ i=1,2,\ldots,N.$$

For $i > m_j$ and $j=2,3,\ldots,M$, compute $G(i,j)$ using the expression

$$G(i,j) = G(i,j-1) + \rho_j G(i-1,j) - \rho_j^{m_j+1} \Phi(i-m_j-1,j) , \qquad (6.4.18)$$

where

$$\Phi(k,j) = \frac{1}{k} \sum_{s=1}^{j-1} \sum_{t=1}^{k} \rho_s^t \, \Phi(k-t,j), \ k=1,2,\ldots,N,$$

and $\Phi(0,j)=1$ for $j=2,3,\ldots,M$. Recursively, calculate $G(N,M)$ that corresponds to the normalizing constant G.

6.5. Other product-form solutions

In this section, we present two cases where we can obtain a product-form solution. The first case deals with a cyclic queueing network with blocking-before-service, and the second case deals with a closed queueing network with blocking consisting of M nodes arbitrarily connected.

A cyclic queueing network with blocking

Consider a closed queueing network consisting of M nodes arranged in tandem as shown in figure 6.2. This network is often referred to as a *cyclic queueing network.* A customer proceeds sequentially through each of these nodes, returning to the first node after it is served at the last node. Let m_i and μ_i be the maximum capacity and the service rate respectively at node i, $i=1,2,\ldots,M$. The service time at each node is assumed to be exponentially distributed. Let N be the total number of customers in the queueing network. We have, $m_i<N$, $i=1,2,\ldots,M$, and $\Sigma m_i>N$. Blocking-before-service mechanism is assumed.

Let $p(\mathbf{n})$ be the probability that the queueing network is in state $\mathbf{n}=(n_1,n_2,\ldots,n_M)$, where n_i is the number of customers in the ith node. Then, the steady-state equations that govern the behaviour of the system are

$$p(\mathbf{n}) \sum_{i=1}^{M} \mu_i \zeta_i(n_i) \eta_{i+1}(n_{i+1}) = \sum_{i=1}^{M} \mu_i \eta_i(n_i) \zeta_{i+1}(n_{i+1}) p(\mathbf{n}+\mathbf{1}_i-\mathbf{1}_{i+1}),$$

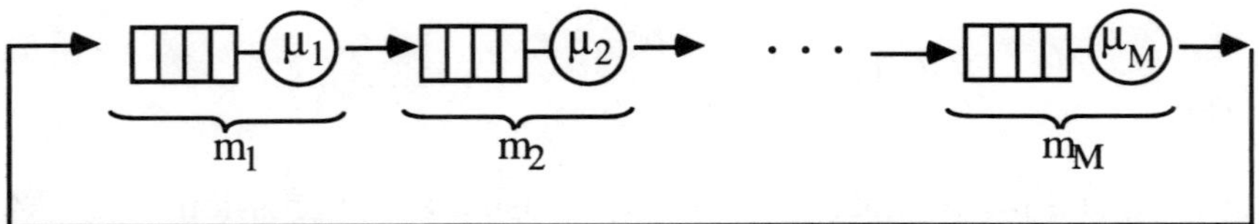

Figure 6.2: A cyclic queueing network with blocking.

where

$$\zeta_i(n_i) = \begin{cases} 1 & \text{if } n_i \neq 0 \\ 0 & \text{if } n_i = 0, \end{cases} \qquad \eta_i(n_i) = \begin{cases} 1 & \text{if } n_i \neq m_i \\ 0 & \text{if } n_i = m_i, \end{cases}$$

and $\Sigma m_i=N$. In general, the steady-state solution for $p(\mathbf{n})$ cannot be obtained for arbitrary M and N. A closed-form solution was given in section 6.3 for M=2. A closed-form solution can be obtained for arbitrary M when the number of customers in the system is such that $N \leq \min_i\{m_i\}$. In this case, no blocking occurs seeing that all N customers fit in the node with the smallest capacity. Therefore, the queueing network has a product-form solution. Another closed-form solution can be obtained for arbitrary M when the number of customers in the system is such that

$$N \geq \sum_{i=1}^{M} m_i - \min_i\{m_i\}.$$

This can be obtained using the concept of *holes*. Let us consider any node of the queueing network, say node i. If there are n_i customers in the node, then n_i spaces are occupied by customers and the remaining m_i-n_i spaces are empty, or we can say that they are occupied by holes. As customers move through the nodes, holes can be seen as moving counter-clockwise. That is, a customer moving from the ith node to the (i+1)st node corresponds to a hole moving from the (i+1)st node to the ith node. It can be shown that the counter-clock movement of holes is also subject to the same blocking mechanism, i.e. blocking-before-service. In view of this, the probability distribution of customers $p(n_1,n_2,...,n_M)$ is equal to the probability distribution of holes $p(m_1-n_1,m_2-n_2,...,m_M-n_M)$ obtained by analyzing the queueing network with N holes and parameters $\{(m_1,\mu_M), (m_M,\mu_{M-1}),...,(m_2,\mu_1)\}$.

Now, when

$$N \geq \sum_{i=1}^{M} m_i - \min_i\{m_i\},$$

the number of holes is Σm_i-N which is less than $\min_i\{m_i\}$. Therefore, the probability distribution of the number of holes can be easily obtained, since the queueing network $\{(m_1,\mu_M),(m_M,\mu_{M-1}),...,(m_2,\mu_1)\}$ has a product-form solution. We have

$$\begin{aligned} p(n_1,n_2,...,n_M) &= p(m_1-n_1,m_2-n_2,...,m_M-n_M) \\ &= \frac{1}{G}\prod_{i=1}^{M}\left(\frac{1}{\mu_i}\right)^{m_{i+1}-n_{i+1}} \\ &= \frac{1}{G'}\prod_{i=1}^{M}\mu_i^{n_{i+1}}, \end{aligned}$$

where G′ is the normalizing constant and $n_{M+1}=n_1$.

A closed queueing network with blocking with $N = \min_i\{m_i\}+1$

Let us consider a closed queueing network with M nodes arbitrarily connected, N customers and blocking-after-service. Each node consists of a single queue served by a server with an exponentially distributed service time with rate μ_i, i=1,2,...,M. A customer upon service completion at node i attempts to enter destination node j with probability r_{ij}, i=1,2,...,M, and j=1,2,...,M. Let b be the node with the smallest buffer, i.e. $m_b=\min_i\{m_i\}$. If $N\leq m_b$, then blocking does not occur and the network has a product-form solution. When N is greater than the minimum buffer capacity, then blocking occurs and product-form solutions are not available. Below, we show that this queueing network has a product-form solution, when N is equal to the minimum buffer capacity plus one, i.e. $N=m_b+1$.

Let $(n_1,n_2,...,n_M)$ be the state vector of this closed queueing network, where n_i is the number of customers in the ith node, $n_i\leq m_i$, i=1,2,...,M, and Σn_i=N. Due to the fact that $N=m_b+1$, the only time that blocking of a node can occur is when node b is full, i.e. it contains m_b customers. Given that this node is full, blocking occurs if the (m_b+1)st customer has just completed its service at an upstream node, say node k, and has chosen to join node b. This state of the network is denoted by the vector

$(0,\ldots,n_k^*=1,\ldots,n_b=m_b,\ldots,0)$, where the * indicates that node k is blocked by node s. For simplicity, it is assumed that $r_{ii}=0$, $i=1,2,\ldots,M$. The steady-state equations are

$$p(\mathbf{n})\sum_{i=1}^{M}\sum_{j=1}^{M} r_{ij}\zeta_i(n_i)\mu_i = \sum_{i=1}^{M}\sum_{j=1}^{M} r_{ij}\eta_j(n_j)\mu_i p(\mathbf{n}+\mathbf{1}_i-\mathbf{1}_j), \mathbf{n}\in S^1 \tag{6.5.1}$$

$$p(0,\ldots,n_k=1,\ldots,n_b=m_b,\ldots,0)r_{kb}\,\mu_k = p(0,\ldots,n_k^*=1,\ldots,n_b=m_b,\ldots,0)\mu_b,$$
$$(0,\ldots,n_k^*=1,\ldots,n_b=m_b,\ldots,0)\in S^2, \tag{6.5.2}$$

where S^1 is the set of states where no node is blocked, S^2 is the set of states where one node is blocked, and $S=S^1\cup S^2$ is the set of all feasible states. The normalizing conditions is

$$\sum_{\mathbf{n}\in S} p(\mathbf{n})=1.$$

Let

$$p(\mathbf{n}) = \begin{cases} q(\mathbf{n}) & \text{if } n_b\le m_b \text{ for } b=1,\ldots,M \\ \dfrac{r_{kb}e_k}{e_b}\, q(0,\ldots,n_b=N,\ldots,0) & \text{if } n_k^*=1 \text{ and } n_b=m_b \end{cases} \tag{6.5.3}$$

where e_i is the visit ratio of the ith node, and $q(\mathbf{n})$ is the solution obtained by assuming that the above queueing network has infinite buffers and m_s+1 customers. $q(\mathbf{n})$ can be easily obtained seeing that in this case the network has a product-form solution. We have

$$q(\mathbf{n}) = \frac{1}{G}\prod_{i=1}^{M}\left(\frac{e_i}{\mu_i}\right)^{n_i},$$

where G is the normalizing constant. By substituting the expression (6.5.3) of $p(\mathbf{n})$ into equations (6.5.1) and (6.5.2), it can be easily verified that the steady-state equations are satisfied. In particular, for states where a node is blocked we have

$$p(0,\ldots,n_k=1,\ldots,n_b=m_b,\ldots,0)r_{kb}\mu_k = p(0,\ldots,n_k^*=1,\ldots,n_b=m_b,\ldots,0)\mu_b,$$

or

$$q(0,...,n_k=1,...,n_b=m_b,...,0)r_{kb}\mu_k = q(0,...,n_b=m_b+1,...,0)\mu_b r_{kb} e_k/e_b,$$

which can be verified by substituting the above product-form solution for q(**n**) given above.

6.6. References

The presentation of the basic concepts on reversible Markov processes in section 6.2 was based on Kelly (1979). The multi-class two-node queueing network with blocking studied in section 6.3 was shown to have a product-form solution by Onvural (1989). In section 6.4, the product-form solution for the closed queueing network with reversible routing is due to Hordijk and van Dijk (1981), the product-form solution for the case of state-dependent routing is due to Yao and Buzacott (1987), and the two special cases which have a product-form solution are due to Hordijk and van Dijk (1981). The multi-class queueing network also presented in section 6.4 was studied by Akyildiz and von Brand (1989a). They obtained a product-form solution which is an extension of the results obtained by Cohen (1979) and van Dijk and Tijms (1986) for a two-node closed queueing network. Finally, the modified recursive procedure for calculating the normalizing constant was proposed by Akyildiz (1989), and it was extended by Akyildiz and von Brand (1989b) to the multi-class case. In section 6.5, the first case is due to Gordon and Newell (1967), and the second case is due to Onvural and Perros (1988/89). Further references related to the topics presented in this Chapter are given below.

Two-node closed queueing networks with blocking

Akyildiz (1987) studied the two-node model discussed in section 6.3 assuming a single class of customers, blocking-after-service and multiple servers. He demonstrated that the queueing network's marginal queue-length distributions are the same as those of an identical queueing network which has infinite capacity queues and a fixed number of customers equal to z-1, where $z=\min\{N,m_1+s_2\}+\min\{N,m_2+s_1\}-N+1$. Balsamo and Donatiello (1989) obtained a recursive technique to compute the cycle time distribution for a two-node closed queueing network with exponential servers and finite capacity queues. Cohen (1979) obtained a product-form solution for a two-node queueing network with blocking assuming multiple classes of customers, class-dependent service time distributions, repetitive-service blocking, and processor sharing with load dependent service times. Class changes were not allowed. Van Dijk and Tijms (1986) showed that the distribution of customers in a two-node queueing network with multiple classes, symmetric scheduling, and repetitive-service blocking is insensitive to the service time distributions. Class changes were not allowed. (see also section 2.5 in Tijms (1986)). Kouvatsos and Almond (1988) analyzed a two-node queueing network with multiple servers and blocking using the maximum entropy principle.

Closed queueing networks with blocking with more than two nodes

The case where the routing probability is state dependent was first analyzed by Kingman (1969), see also section 6.1 in Kelly (1979). Other cases of single class closed queueing networks with blocking which have product-form solutions have been investigated by Hordijk and van Dijk (1983), Yao and Buzacott (1987), van Dijk (1989), and Akyildiz and von Brand (1989d). Balsamo and Iazeolla (1983) demonstrated that certain queueing networks with blocking with a product-form solution, are equivalent to queueing networks with infinite capacity queues. Two queueing networks are considered equivalent if they have the same steady-state probabilities. The authors obtained equivalencies for the queueing network with blocking under the assumption of a) reversible routing and b) constant service rate. Such equivalencies allow the study of queueing networks with blocking using conventional computational techniques. Balsamo, Clò and Donatiello (1993) obtained the cycle time distribution of cyclic queueing networks with blocking with or without a product-form solution. Also, Balsamo and Clò (1992) obtained a relationship between the joint queue-length distribution at arrival instants and at arbitrary instants for closed exponential queueing networks with a general topology and various blocking types. In the special case of product-form cyclic queueing networks with blocking and central server networks with blocking, they showed that the state distribution at arrival instants is identical to the state distribution at arbitrary times of the queueing network with one less customer and modified finite buffer capacities.

Le Ny (1980) considered a configuration of the *central server* type with multiple classes. The configuration consists of a node (call it node 0) linked to M nodes in parallel. Each of these M nodes is linked to node 0. Change of the class of a customer is not permitted. The routing probability from node 0 to any of the M nodes is state-dependent and it is given by expression (6.4.7). The rate of departure from node i, for class k customers, has the form $h_i(n_i)h_{ik}(n_{ik})$, where $h_i(n_i)$ is a function of the number of customers n_i in the node i, and $h_{ik}(n_{ik})$ is a function of the number of class k customers n_{ik} in node i. Le Ny obtained a product-form solution for this particular queueing network using the notion of *exchangeable station.* This notion is due to Pellaumail (see Le Ny (1980) and (1983)). Other studies of queueing networks with blocking and multiple classes which have a product-form solution can be found in Lazar and Robertazzi (1984), Pittel (1979), Yao and Buzacott (1985), Akyildiz and von Brand (1989c), Akyildiz and van Dijk(1990).

A survey of queueing networks with blocking and their equivalencies was compiled by Balsamo and De Nitto Personè (1992). Finally, a survey of the literature related to closed queueing networks with blocking can be found in Onvural (1990).

Bibliography

Akyildiz, I. F., Exact product form solution for queueing networks with blocking, *IEEE Trans. Comp.* **36** (1987) 122-126.

Akyildiz, I. F., Exact analysis of queueing networks with rejection blocking, in: Perros and Altiok, (Eds.), *Proc. First International Workshop on Queueing Networks with Blocking*, (North-Holland, Amsterdam, 1989) 19-29.

Akyildiz, I. F. and von Brand, H., Exact solutions for open, closed and mixed queueing networks with rejection blocking, *Theoret. Comput. Sci.* **64** (1989a) 203-219.

Akyildiz, I. F. and von Brand, H., Computational algorithms for networks of queues with rejection blocking, *Acta Inf.* **26** (1989b) 559-576.

Akyildiz, I. F. and von Brand, H., Central server models with multiple job classes, state dependent routing, and rejection blocking, *IEEE Trans. Soft. Eng.* **15** (1989c) 1305-1312.

Akyildiz, I. F. and von Brand, H., Dual and self dual networks of queues with rejection blocking, *Computing J.* **43** (1989d) 1-11.

Akyildiz, I. F. and van Dijk, N. M., Exact solution of parallel queues with finite buffers, in: King, Mitrani, and Pooley, (Eds.), *Performance '90* (North-Holland, Amsterdam, 1990) 3549.

Balsamo, S. and Clò, M. C., *State distribution at arrival times for closed queueing networks with blocking*, Tech. Rept. TR-35/92, Dipartimento di Informatica, Universita di Pisa, 1992.

Balsamo, S., Clò, M. C., and Donatiello, L., Cycle time distribution of cyclic networks with blocking, in: Onvural and Akyildiz, (Eds.), *Proc. Second International Workshop on Queueing Networks with Finite Capacity*, (North-Holland, Amsterdam, 1993) 27-40.

Balsamo, S. and Donatiello, L., On the cycle time distribution in a two-stage queueing network with blocking, *IEEE Trans. Soft. Eng.* **15** (1989) 1206-1216.

Balsamo, S. and Iazeolla, G., Some equivalence properties for queueing networks with and without blocking, in: Agrawala and Tripathi, (Eds.), *Performance '83*, (North-Holland, Amsterdam, 1983), 351-360.

Balsamo, S. and De Nitto Personè, V., *A survey of product-form queueing networks with blocking and their equivalences*, Tech. Rept., Dipartmento di Ingegneria Elettronica, Universita di Roma, 1992. To appear in *Ann. Oper. Res.*

Barbour, A. D., Networks of queues and the method of stages, *Adv. Appl. Probability* **8** (1976) 249-260.

Baskett, F., Chandy, K., Muntz, R. R. and Palacios, F. G., Open closed and mixed networks of queues with different classes of customers, *J. ACM* **22** (1975) 249-260.

Cohen, J. W., The multiple phase service network with generalized processor sharing, *Acta Infor.* **12** (1979) 245-284.

Gelenbe, E. and Mitrani, I., *Analysis and Synthesis of Computer Systems*, (Academic Press, 1980)

Gordon, W. J. and Newell, G. F., Cyclic queueing systems with restricted length queues, *Oper. Res.* **15** (1967) 266-278.

Hordijk, A. and van Dijk, N., Networks of queues with blocking, in: Kylstra, (Ed.), *Performance '81*, (North-Holland, Amsterdam, 1981) 51-65.

Hordijk, A. and van Dijk, N., Networks of queues: Part I: Job-local-balance and the adjoint process. Part II: General routing and service characteristics, in: Baccelli and Fayolle, (Eds.), *Modelling and Performance Evaluation Methodology*, (Springer-Verlag, Lecture Notes in Control and Information Science, Vol 60, 1983) 158-205.

Kelly, F. P., *Reversibility and stochastic networks*, (Wiley, 1979).

Kingman, J. F. C., Markov population processes, *J. Appl. Prob.* **6** (1969) 1-18.

Kouvatsos, D. D. and Almond, J., Maximum entropy two-station cyclic queues with multiple general servers, *Acta Inf.* **26** (1988) 241-267.

Lazar, A. A. and Robertazzi, T. G., *The geometry of lattices for Markovian queueing networks*, Tech. Rept., Electrical Engineering Dept., Columbia Univ., 1984.

Le Ny, L.-M., Étude analytique de reseaux de files d'attende multiclasses à routage variable, *RAIRO Recherche Operationelle* **14** (1980) 331-347.

Le Ny, L.-M., Forme produit pour des reseaux multiclasses à routage dynamique, *Annales Scientifiques de l'Universitè de Clermont-Ferrand II* **76** (1983) 17-34.

Onvural, R. O., A note on the product from solutions of multiclass closed queueing networks with blocking, *Perf. Eval.* **10** (1989) 247-253.

Onvural, R. O., Survey of closed queueing networks with blocking, *ACM Comput. Surveys* **22** (1990) 83-121.

Onvural, R. O. and Perros, H. G., Equivalencies between open and closed queueing networks with finite buffers, *Perform. Eval.* **9** (1988/89) 263-269.

Pittel, B., Closed exponential networks of queues with saturation: The Jackson-type stationary distribution and its asymptotic analysis, *Math. Oper. Res.* 4 (1979) 367-378.

Tijms, H. C., *Stochastic modelling and analysis, a computational approach*, (Wiley, 1986).

van Dijk, N. M., Comment on Yao and Buzacott's "modeling a class of flexible manufacturing systems with reversible routing", *Oper. Res.* 37 (1989) 845-846.

van Dijk, N. M. and Tijms, H. C., Insensitivity in two-node blocking models with applications, in: Boxma, Cohen and Tijms, (Eds.), *Teletraffic Analysis and Computer Performance Evaluation*, (North-Holland, Amsterdam, 1986) 329-340.

Yao, D. D. and Buzacott, J. A., Modeling a class of state-dependent routing in flexible manufacturing systems, *Ann. Oper. Res.* **3** (1985) 153-167.

Yao, D. D. and Buzacott, J. A., Modeling a class of flexible manufacturing systems with reversible routing, *Oper. Res.* **35** (1987) 87-93.

CHAPTER 7

Closed Queueing Networks With Blocking With Non Product-Form Solution

7.1. Introduction

In this Chapter, we examine closed queueing networks with blocking that do not have a product-form solution. In section 7.2, we describe some properties of closed queueing networks with blocking, and in section 7.3, we show that the throughput of a cyclic queueing network with blocking-before-service is symmetrical with respect to the total number of customers in the network. In section 7.4, we give an approximation algorithm for estimating the throughput based on the results presented in sections 7.2 and 7.3. In section 7.5, we analyze a special case of cyclic queueing networks where all nodes are indistinguishable. In section 7.6, we describe three approximation algorithms which have been proposed for the analysis of arbitrary configurations of closed queueing networks with blocking. Finally, relevant references are given in section 7.7.

7.2. Some properties of closed queueing networks with blocking

In this section, we identify certain conditions under which the rate matrix of a closed queueing network with blocking remains the same for a range of values of the total number of customers in the network.

Consider a closed queueing network with blocking consisting of M nodes and N customers. Each node consists of a single server finite capacity queue. The service time at the ith node is exponentially distributed with rate μ_i. Let m_i denote the capacity of the ith node. A customer upon completion of its service at node i, i=1,2,...,M, attempts to enter the jth node with probability r_{ij}, j=1,2,...,M, where $\sum_j r_{ij}=1$. Blocking-after-service is assumed. Due to the nature of this blocking mechanism and due to the fact that the finite capacity nodes are arbitrarily interconnected, it is possible that deadlocks may occur. For instance, assume that node i is blocked by node j. Now, it is possible that a customer in node j, upon completion of its service, may choose to go to node i. Deadlock will occur if node i is full at that moment. It is assumed that deadlocks are detected and resolved instantaneously. In particular, they are resolved by instantaneously exchanging the blocked customers. In general, this resolution may violate the first-blocked-first-enter priority rule associated with blocking-after-service. Let B denote the sum of the capacities of all the nodes, that is $B=\sum m_i$. Also, let ℓ be the node with the largest capacity, that is, $m_\ell=\max_i\{m_i\}$.

Theorem 1: Consider the above queueing network. If $m_\ell>B-m_\ell$, then the queueing network has the same rate matrix for all N such that $B-m_\ell+1 \leq N\leq m_\ell$.

Proof: Consider all nodes other than node ℓ. We first note that when $B-m_\ell+1\leq N\leq m_\ell$, the distribution of the N customers over the remaining M-1 nodes is independent of N. This is because N takes values greater than $B-m_\ell$, the total capacity of the remaining M-1 nodes. (Obviously, this will not be the case if $N<B-m_\ell$.) In order to clarify this point, let us consider the states of the system for $N>B-m_\ell$. There is one state in which all the remaining M-1 nodes are full, and node ℓ contains the left-over $N-(B-m_\ell)$ customers. There is another state in which all the remaining M-1 nodes are empty, and node ℓ contains all N customers. The remaining states reflect all possible combinations in between these two states, with

$$n_\ell = N-\sum_{i\neq\ell} n_i,$$

where n_i is the number of customers in node i. Now, consider another population of customers $N'\geq N$, where $N'\leq m_\ell$. Then, we can easily verify that for N' we will obtain the same states, only node ℓ will contain $N'-N$ customers more. It can be also verified that the number of states where one or more nodes are blocked, is independent of N as long as $N\geq B-m_\ell$. Furthermore, for $B-m_\ell+1\leq N\leq m_\ell$, node ℓ cannot block any other node and it cannot be empty. Hence transitions into and out of node ℓ are independent of N. Therefore, for all $B-m_\ell\leq N\leq m_\ell$ we have the same rate matrix. That is, for N and N' such that $B-m_\ell\leq N\leq N'\leq m_\ell$ we have

$$p^N(n_1,\ldots,n_\ell,\ldots,n_M) = p^{N'}(n_1,\ldots,n_\ell+N'-N,\ldots,n_M),$$

where $p^N(.)$ and $p^{N'}(.)$ are the steady-state queue-length distribution with N and N′ customers in the network respectively.□

We demonstrate this theorem through an example. Consider a three-node queueing network with blocking-after-service, and let $m_1=12$, $m_2=5$, and $m_3=3$. The total capacity B of the network is 20, and the node with the largest capacity is node 1, that is $\ell=1$. The queueing network has the same rate matrix for N=9,10,11,12. Also, for N and N′ such that $9 \le N \le N' \le 12$,

$$p^N(n_1,n_2,n_3) = p^{N'}(n_1+N-N',n_2,n_3).$$

For instance, $p^9(3,4,2)=p^{10}(4,4,2)=p^{11}(5,4,2)=p^{12}(6,4,2)$. Let $L_i(N)$ be the mean number of customers in node i when the population in the network is N. Then, we have that for N=10,11,12, $L_1(N)=1+L_1(N-1)$, and for N=9,10,11,12, $L_2(N)$ and $L_3(N)$ remain the same. Finally, the network's throughput remains the same for N=9,10,11,12.

In addition to the range of values $B-m_\ell+1 \le N \le m_\ell$, for which the above queueing network has the same rate matrix, the network has also the same rate matrix when $N=m_\ell+1$ given that some of its states are properly aggregated. When $N=m_\ell+1$, there can be at most one node that can be blocked by node ℓ at any time. In this case, the blocked node has exactly one customer (which has completed its service), node ℓ is full and all the other nodes are empty. The state in which node k is blocked by node ℓ is denoted by $(0,\ldots,\overset{*}{n_k}=1,\ldots,\overset{k}{n_\ell}=m_\ell,\ldots,0)$, where the superscripts * and k indicate that node k is blocked and node ℓ is blocking node k respectively. The global balance equation of this state is

$$p(0,\ldots,\overset{*}{n_k}=1,\ldots,\overset{k}{n_\ell}=m_\ell,\ldots,0)\mu_\ell$$

$$= p(0,\ldots,n_k=1,\ldots,n_\ell=m_\ell,\ldots,0)\mu_k r_{k\ell}. \qquad (7.2.1)$$

The left-hand side of the above equation is the rate out of the state where node k is blocked by node ℓ , and the right-hand side is the rate into this state. Now, let us define the state $(0,\ldots,n_\ell=m_\ell+1,\ldots,0)$ such that

$$p(0,\ldots,n_\ell=m_\ell+1,\ldots,0) = \sum_{k=1,k\neq\ell}^{M} p(0,\ldots,\overset{*}{n_k}=1,\ldots,\overset{k}{n_\ell}=m_\ell,\ldots,0). \qquad (7.2.2)$$

That is, $p(0,\ldots,n_\ell=m_\ell+1,\ldots,0)$ is the probability that node ℓ is blocking a node. Summing both sides of (7.2.1) over k, $k\neq\ell$, and using (7.2.2) we have

$$p(0,\ldots,n_\ell=m_\ell+1,\ldots,0)\mu_\ell = \sum_{k=1,k\neq\ell}^{M} p(0,\ldots,n_k=1,\ldots,n_\ell=m_\ell,\ldots,0)\mu_k r_{k\ell}.$$

Furthermore, a departure from a state where node ℓ is blocking node k is independent of the blocked node. This is because all nodes other than node ℓ are empty and (i) if $m_k>1$, then there is a space available in node k, or (ii) if $m_k=1$, then a deadlock will occur which will be detected immediately and will be resolved by exchanging the blocked customers. Therefore, a customer at node ℓ joins any node k with rate $r_{k\ell}\mu_k$. Now, it can be easily shown that if we aggregate all the states where node ℓ is blocking a node to the single state $(0,\ldots,n_\ell=m_\ell+1,\ldots,0)$, the resulting rate matrix is identical to the rate matrix of the queueing network when $B-m_\ell+1 \leq N\leq m_\ell$.

We now consider the same queueing network with N customers such that the node with the largest capacity, node ℓ, cannot be empty. We will show below that if the buffer capacity m_ℓ of this node and the number of customers N are increased by the same amount, then the rate matrix of the queueing network does not change.

Theorem 2: Consider the queueing network described at the beginning of this section. Also, consider a similar queueing network with the same parameters except that the buffer capacity of node ℓ is increased to m'_ℓ. Let N be the number of customers in the first queueing network, and let N be such that $N\geq B-m_\ell+1$. That is, node ℓ is never empty. Then, the first queueing network with N customers has the same rate matrix as the second queueing network with $N'=N+m'_\ell-m_\ell$.

Proof: Let n_ℓ and n'_ℓ be the number of customers in node ℓ in the first queueing network and in the second queueing network respectively. We first note that $N-(B-m_\ell)\leq n_\ell\leq \min\{m_\ell,N\}$ in the first queueing network, and $N'-(B'-m'_\ell)\leq n'_\ell\leq \min\{m'_\ell,N'\}$ in the second queueing network. Hence, n_ℓ and n'_ℓ take the same number of (different) values. Furthermore, given the number of customers n_ℓ in the first queueing network and $n'_\ell=n_\ell+N'-N$ in the second queueing network, node ℓ in both networks cannot be empty, and the number of customers left to be distributed over the remaining M-1 nodes are the same in both networks.

Similarly, it can be argued that the number of states in which one or more nodes are blocked is the same in both networks. (This is not the case if node ℓ is allowed to be empty.) Hence, both networks have the same number of states. Furthermore, state $(n_1,...,n_\ell,...,n_M)$ of the first node is equivalent to state $(n_1,...,n_\ell+N'-N,...,n_M)$ of the second network in the sense that both states have the same transition rates into and out of corresponding equivalent states. Hence,

$$p^{QN1}(n_1,...,n_\ell,...,n_M) = p^{QN2}(n_1,...,n_\ell+N'-N,...,n_M),$$

where $p^{QN1}(.)$ and $p^{QN2}(.)$ are the steady-state queue-length distribution of the first and second queueing network respectively.□

As an example, let the first closed queueing network be a three-node network with $m_1=4$, $m_2=3$, $m_3=2$ and $N=6$ customers. Let the second queueing network be identical to the first one with $m_1'=5$, $m_2=3$, $m_3=2$ and $N'=7$. Then, the two networks have the same rate matrix. Hence,

$$p^{QN1}(n_1,n_2,n_3) = p^{QN2}(n_1+1,n_2,n_3).$$

For instance,

$$p^{QN1}(1,2,3) = p^{QN2}(2,2,3),$$

$$p^{QN1}(4,1,1) = p^{QN2}(5,1,1).$$

Also,

$$L_1^{QN2} = 1+L_1^{QN1},\ L_2^{QN2} = L_2^{QN1},\ \text{and}\ L_3^{QN2} = L_3^{QN1},$$

where L_i^{QNj} is the mean number of customers in the ith node of the jth queueing network. Finally, the throughput of node i in the first queueing network is the same as the throughput of node i in the second queueing network, i=1,2,3.

We now consider the same queueing network examined above but assuming blocking-before-service. As before, it is possible that a deadlock may occur. For instance, assume that node i is blocked by node j. That is, a customer at node i has chosen node j as its destination, and node j is full. Now, it is possible that a customer in node j, prior to the beginning of its service, chooses node i as its destination. Deadlock will occur if node i is full at that moment. The theorem is an extension of theorem 1, and it can be proved following similar arguments.

Theorem 3: Consider a deadlock free queueing network under blocking-before-service. Then, if $m_\ell > B - m_\ell$, the queueing network has the same rate matrix for $B - m_\ell \leq N \leq m_\ell$.

7.3. The throughput of cyclic queueing networks with blocking-before-service

Let us consider a cyclic queueing network consisting of M nodes arranged in tandem as shown in figure 7.1. A customer proceeds sequentially through each of these nodes, returning to the first node after it is served at the last node. Let m_i be the capacity of node i, i=1,2,...,M, B the sum of all nodal capacities, i. e. $B = \sum m_i$, and N the total number of customers in the queueing network. We have, $m_i < N$, i=1,2,...,M, and $B > N$. We assume blocking-before-service and phase-type distributed service times at each server. We will denote this cyclic queueing network as QN.

Let $T_i(N)$ and $T(N)$ denote respectively the throughput of node i and the throughput of the network when there are N customers in the network. We have, $T_i(N) = T(N)$, i=1,2,...M. In this section, we will show that the throughput of this cyclic queueing network is symmetrical with respect to N. That is

$$T(N) = T(B-N), \quad N=0,1,\ldots,B.$$

We first note that the symmetry property is satisfied when N=0 or N=B, seeing that in both cases the throughput is zero. (When N=B, all nodes are full. Due to the blocking-before-service mechanism, no service will ever begin and consequently the throughput is zero.) Therefore, we will restrict the proof to the case where $0 < N < B$. The proof is based on the comparison of sample paths of the network with populations N and B-N. By convention, we will assume that if i=M then node i+1 is node 1, and if i=1 then node i-1 is node M.

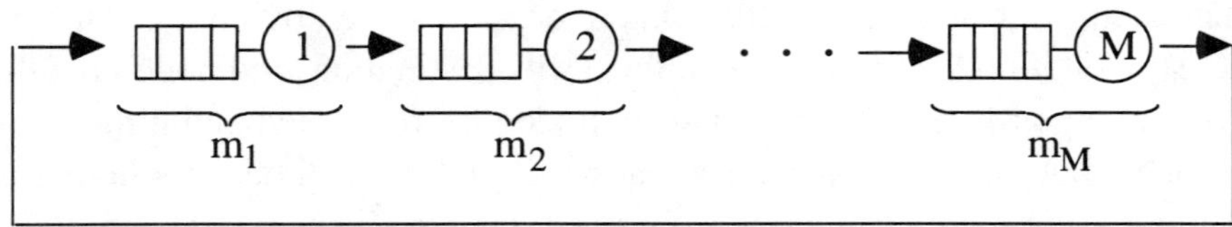

Figure 7.1: A cyclic queueing network with blocking.

The initial condition of the queueing network at time t=0 is defined by the vector $\mathbf{k}=(k_1,k_2,...,k_M)$, where k_i is the number of customers in the ith node. The vector **k** will be referred to as the *initial* marking. Also, let $\mathbf{h}=(h_1,h_2,...,h_M)$ be a vector where h_i indicates the number of holes (free spaces) in node i at time t=0. Obviously, for each node i we have $k_i+h_i=m_i$. The queueing network associated with an initial marking **k** will be denoted as (QN,**k**). We assume that all servers which are neither idle nor blocked for this initial marking initiate a service at time t=0.

The duration of the service periods at the ith node are given by a sequence of non-negative service times $\{t_{i,n}\}$, $n\geq1$, $i=1,2,...,M$. Let $C_{i,n}$, $n\geq1$, denote the time of the beginning of the nth service at the ith server. We assume that $C_{i,1}=0$ for all servers which are neither idle nor blocked at time t=0. Also, let $D_{i,n}$, $n\geq1$, denote the time of the nth service completion at the ith server. Since we assume blocking-before-service, the service completion instant is also the departure instant of the customer. The quantities $t_{i,n}$, $C_{i,n}$ and $D_{i,n}$ are related by the following two evolution equations.

Lemma 1: The evolution equations for (QN,**k**) are

$$D_{i,n} = C_{i,n}+t_{i,n}\ ,\ n\geq1, \tag{7.3.1}$$

$$C_{i,n} = \max\{D_{i,n-1},\ D_{i-1,n-k_i},\ D_{i+1,n-h_{i+1}}\},\ n\geq1,\ 1\leq i\leq M, \tag{7.3.2}$$

where $D_{i,n}=0$, $n<0$, and $i=1,2,...M$.

Proof: Equation (7.3.1) simply states that the time of the completion of the nth service is equal to the time of its beginning plus the duration of the service time. Now, let us consider equation (7.3.2). In order for the nth service to begin, the following conditions must hold:

1. The (n-1)st service has to be completed. This happens at time $D_{i,n-1}$.
2. At least n arrivals to the ith node have to take place, in order to guarantee that the ith node is not empty. Seeing that the node may have started with an initial marking k_i, we have that at least $n-k_i$ arrivals have to occur. The time of the $n-k_i$ arrival is the departure time of the $n-k_i$ customer from the previous node, i.e. $D_{i-1,n-k_i}$.
3. There should be at least one space free in the downstream node. That is, at least n departures, or more specifically $n-h_{i+1}$ departures if the node started with h_{i+1} holes. The time for the $n-h_{i+1}$ departure is $D_{i+1,n-h_{i+1}}$.

Since all three conditions have to hold, the instant at which the nth service begins is the maximum of the three times $D_{i,n-1}$, $D_{i-1,n-k_i}$, and $D_{i+1,n-h_{i+1}}$.□

Now let D_n be the time at which n service completions have occurred at each server. This time will be referred to as the nth overall completion time. We have

$$D_n = \max_i\{D_{i,n}\},\ n\geq 1. \qquad (7.3.3)$$

Suppose that we are interested in the time of the last n_0 service completions at all servers, i.e. D_{n_0}, $n_0\geq 1$. This quantity can be expressed as the length of a longest path of a directed graph. This graph will be referred to as the *activity graph* of the network, and it will be denoted by G_{n_0}. The nodes of the graph represent the times of beginning and completion of services and the arcs represent the timing constraints as expressed by the evolution equations (7.3.1) and (7.3.2). Each node is denoted by $C_{i,n}$ (or $D_{i,n}$) and it represents the time that the nth service begins (or ends). Also, there are two additional nodes, namely C_1 and D_{n_0}. Node C_1 is associated with the initial conditions, i.e. the system starts at time t=0, and node D_{n_0} is associated with the instant of the overall completion time D_{n_0}. There is one arc from node $C_{i,n}$ to node $D_{i,n}$ which represents a service activity and its associated weight is the corresponding service time $t_{i,n}$. This arc expresses the relationship given by (7.3.1). For each node $C_{i,n}$ there are, in general, three arcs oriented towards the node. These three arcs represent the three components of the right-hand side of (7.3.2) and their associated weight is zero. Also, there is an arc from C_1 to each $C_{i,1}$, i=1,2,...,M, with zero associated weight. These arcs ensure that the servers can start working at time t=0. Finally, there is an arc with zero weight between each node D_{i,n_0} and node D_{n_0}. These arcs represent the relationship given by (7.3.3).

As an example, consider the three-node cyclic queueing network shown in figure 7.2. The buffer capacities are m_1=2, m_2=1, m_3=2, the initial marking is **k**=(1,0,2), and the initial holes are **h**=(1,1,0). The activity graph of this queueing network with the initial marking **k**=(1,0,2) for n_0=3 is given in figure 7.3.

We now consider the same queueing network QN with B-N customers and with a *symmetrical* initial marking $\mathbf{k}^s=(k_1^s,k_2^s,...,k_M^s)$, where $k_i^s=h_i$. Let (QN,$\mathbf{k}^s$) denote the network with this initial marking. The symmetrical

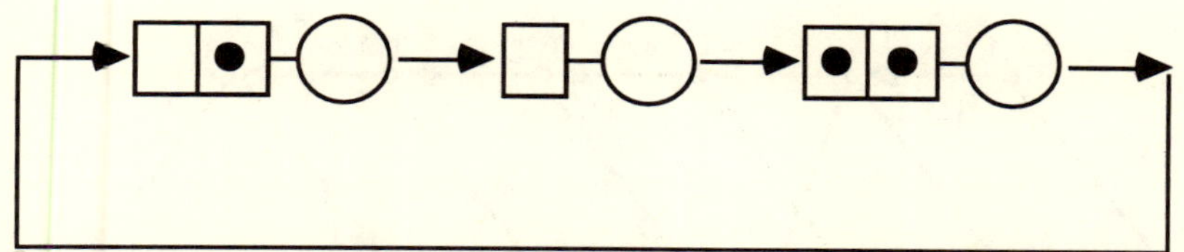

Figure 7.2: Three-node cyclic queueing network with k=(1,0,2). (Adopted from Dallery and Towsley (1991))

initial marking corresponds to the initial marking of holes in (QN,**k**) and vice-versa. That is $\mathbf{k}^s=\mathbf{h}$, and $\mathbf{h}^s=\mathbf{k}$. The symmetrical initial marking of the queueing network shown in figure 7.2 is (1,1,0).

Let $C^s_{i,n}$, $D^s_{i,n}$, and $t^s_{i,n}$ denote the equivalent quantities for $C_{i,n}$, $D_{i,n}$ and $t_{i,n}$ in (QN,$\mathbf{k}^s$). The evolution equations of (QN,$\mathbf{k}^s$) are similar to those of (QN,**k**) given by (7.3.1) and (7.3.2). Also, the activity graph $G^s_{n_0}$ of (QN,$\mathbf{k}^s$) can be constructed in the same way as G_{n_0} of (QN,**k**). These two activity graphs are related as follows.

Lemma 2: Assume that the sequences of service times in (QN,$\mathbf{k}^s$) are the reverse of those in (QN,**k**), i.e. $t^s_{i,n}=t_{i,n_0+1-n}$, i=1,2,...M, n=1,2,...,n_0. Then, $G^s_{n_0}$ of (QN,$\mathbf{k}^s$) can be obtained from G_{n_0} of (QN,**k**) by reversing its arcs.

Proof: This can be shown by establishing a one-to-one correspondence between the arcs of G_{n_0} and $G^s_{n_0}$. Define the following correspondences:

$$D^s_{i,n_0+1-n} \leftrightarrow C_{i,n},\ i=1,2,...,M,\ n=1,2,...,n_0$$

$$C^s_{i,n_0+1-n} \leftrightarrow D_{i,n},\ i=1,2,...,M,\ n=1,2,...,n_0$$

$$D^s_{n_0} \leftrightarrow C_i$$

$$C^s_i \leftrightarrow D_{n_0}.$$

It is easy to check that for any arc connecting two nodes in G_{n_0} there is an arc connecting the two corresponding nodes in $G^s_{n_0}$. This arc has the same weight but it is oriented in the reverse direction. □

We can now state the following theorem.

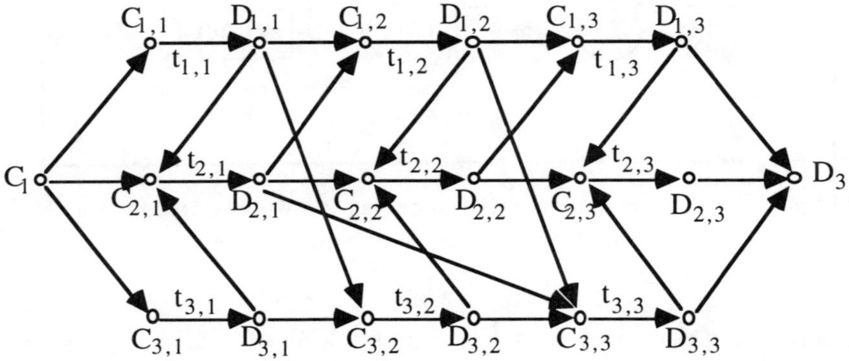

Figure 7.3: Activity graph G_3.
(Adopted from Dallery and Towsley (1991))

Theorem 1: Under the assumption that the service times of all servers are i.i.d. random variables, the overall completion times D_{n_0} in (QN,**k**) and $D^s_{n_0}$ in (QN,$\mathbf{k}^s$) have the same distribution for all $n_0 \geq 1$.

Proof: Since $G^s_{n_0}$ is the reverse of G_{n_0}, the length of the longest path $D^s_{n_0}$ is equal to the length of the longest path D_{n_0}. Now, assuming that the service times are id. the sequence of service times $\{t^s_{i,n}\}$, $n=1,2,...,n_0$, has the same joint distribution as the sequence $\{t_{i,n}\}$, $n=1,2,...,n_0$. Therefore, the overall completion times D_{n_0} in (QN,**k**) and $D^s_{n_0}$ in (QN,$\mathbf{k}^s$) have the same distribution for all $n_0 \geq 1$. □

Let us now restrict our attention to phase-type distributions. In this case, the cyclic queueing network can be described by an irreducible state Markov process. Consequently, its stationary probability distribution is independent of the initial state. Thus, the asymptotic throughput T(N) exists and it depends only on the total number of customers N, and it is otherwise independent of the initial marking. Thus, T(N) can be expressed as

$$T(N) = \lim_{n_0 \to \infty} \left[\frac{E(D_{n_0})}{n_0} \right]^{-1} .$$

In view of the above definition of throughput and theorem 1, it follows that the throughput T(N) is symmetric with respect to the total number of customers N.

We conclude this section by examining the *reverse* and *dual* cyclic queueing network of QN. Let QN^r be the reverse cyclic queueing network obtained from QN. In this network customers proceed in the opposite direction than in QN. That is, customers in node i are served by the server of node i-1 and upon service completion they move to the buffer of node i-1. Blocking-before-service is assumed. Let $T^r(N)$ be the throughput of the reverse queueing network QN^r. Then, it can be shown that its throughput is the same as that of QN, i.e. $T^r(N)=T(N)$, N=0,1,...,B. This can be proved following similar arguments as above. In particular, consider QN^r with the same initial marking as QN, i.e. $\mathbf{k}^r=\mathbf{k}$. Then, the evolution equations of $(QN^r,\mathbf{k}^r)$ can be shown to be the same as those of $(QN,\mathbf{k}^s)$ provided that $t^r_{i,n}=t^s_{i,n}$, i=1,2,...,M, n≥1. In view of this, $(QN^r,\mathbf{k}^r)$ and $(QN,\mathbf{k}^s)$ have the same activity graph. Consequently, for the same total number of customers N, QN^r and QN have the same throughput.

Now, let QN^d be the dual cyclic queueing network of QN. This network is obtained by reversing the flow of customers as in the reverse cyclic queueing network QN^r and assuming B-N customers. Blocking-before-service is assumed. Let $\mathbf{k}^d$ be the initial marking of QN^d, where $\mathbf{k}^d=\mathbf{h}$. The behaviour of customers in the dual network is the same as the behaviour of holes in QN described in section 6.5. It is easy to show that $(QN^d,\mathbf{k}^d)$ is the dual of $(QN,\mathbf{k}^s)$ and that their stochastic behaviour is equivalent.

7.4. Approximate throughput analysis of cyclic queueing networks with blocking

In this section, we present an approximation algorithm for calculating the throughput curve of a cyclic queueing network with blocking-after-service or blocking-before-service. The algorithm is based on the results obtained in the previous two sections. As before, we assume that the cyclic queueing network consists of M nodes, m_i is the maximum capacity of node i, i=1,2,...,M, B is the sum of all nodal capacities, i.e. $B=\sum m_i$, and N is the total number of customers in the queueing network, where $m_i<N$ and $B>N$. Let b and ℓ indicate the node with the smallest respectively largest buffer, i.e. $m_b=\min_i\{m_i\}$ and $m_\ell=\max_i\{m_i\}$. The service time at the ith node is assumed to be exponentially distributed with rate μ_i, i=1,2,...,M. Let T(N) be the throughput of the queueing network. This is identical to the throughput of any node in the cyclic network. The approximation algorithm presented in this section does not give the throughput for a given value of N. Rather, it gives the entire throughput curve obtained by varying N. This is done by fitting a curve that passes through a number of known points. Below, we first describe the approximation algorithm for blocking-before-service and then for blocking-after-service.

Blocking-before-service

As it was shown in the previous section, the throughput curve obtained by varying N is symmetrical with respect to N. In addition, the following theorem stated without proof establishes the monotonicity of the throughput for a specific range of values of N.

Theorem 1: Consider a cyclic queueing network with blocking-before-service and exponentially distributed service times. Then for $0<N<m_\ell$, we have that $T(N) \le T(N+1)$.

In view of these results, the throughput curve has the following shape. The maximum of the curve occurs at the point $N=B/2$ if B is even, and at points $\lfloor B/2 \rfloor$ and $\lfloor B/2 \rfloor+1$ if B is odd. T(N) is non-decreasing as N increases from 1 to B/2 (or $\lfloor B/2 \rfloor$) and non-increasing as N increases from B/2 (or $\lfloor B/2 \rfloor+1$) to B-1.

From theorem 3 given in section 7.2 we have that if $m_\ell > B-m_\ell$, then the queueing network has the same rate matrix for $B-m_\ell \le N \le m_\ell$. Therefore, for this range of values of N the throughput of the network will be the same. Furthermore, since we know where the maximum occurs, and using the fact that the mid-point(s) of the values in the set $[B-m_\ell, m_\ell]$ is B/2 if B is even and $\lfloor B/2 \rfloor$ and $\lfloor B/2 \rfloor+1$ if B is odd, we conclude that the throughput is maximum for all $B-m_\ell \le N \le m_\ell$. As an example, consider a cyclic network with B=12 and $m_\ell=8$. Then, the rate matrix will be identical for $4 \le N \le 8$. Therefore, T(4)=T(5)=...=T(8). Also, the maximum will occur at N=6, which means that the maximum of this throughput curve occurs at N=4,5,...,8. This throughput curve is shown in figure 7.4.

Therefore, if the above condition applies, we fit a curve through the points $T(1),...,T(B-m_\ell)$. Otherwise, we fit a curve through the points T(1) to $T(\lfloor B/2 \rfloor)$. Now, the points $T(1),...,T(m_b)$ can be easily determined since at these points the network has a product-form solution. The maximum throughput $T(N^*)$ is calculated by solving the cyclic queueing network numerically using the techniques described in Chapter 2 for $N^*=B-m_\lambda$ or $N^*=\lfloor B/2 \rfloor$ (depending upon which one is applicable). The fitted curve has the form

$$T(N) = T(N+1) - \alpha\chi^{N^*-N}. \qquad (7.4.1)$$

This expression takes into account the rate at which the throughput changes as the number of customers change. α can be recursively determined as follows. For $N=m_b$, we have

$$\begin{aligned} T(m_b) &= T(m_b+1) - \alpha\chi^{N^*-m_b} \\ &= T(m_b+2) - \alpha\chi^{N^*-m_b-1} - \alpha\chi^{N^*-m_b} \\ &\vdots \\ &= T(N^*) - \alpha \sum_{i=1}^{N^*-m_b} \chi^i. \end{aligned}$$

Hence,

$$\alpha = \frac{T(N^*) - T(m_b)}{\sum_{i=1}^{N^*-m_b} \chi^i}. \qquad (7.4.2)$$

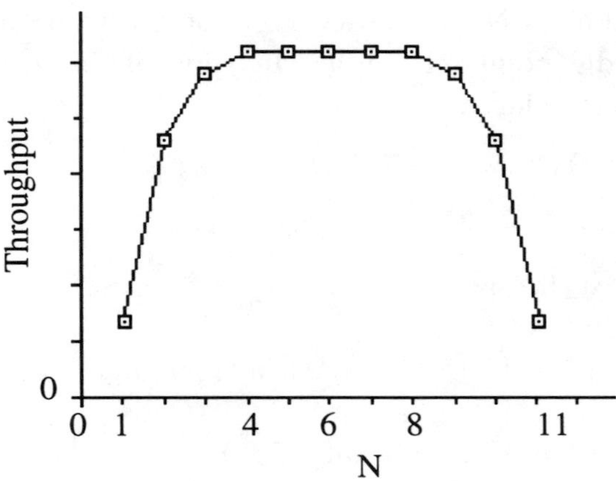

Figure 7.4: Throughput vs N (blocking-before-service).

Substituting α into (7.4.1) gives

$$T(N) = T(N+1) - \chi^{N^*-N} \frac{T(N^*) - T(m_b)}{\sum_{i=1}^{N^*-m_b} \chi^i} .$$

Now, for $N=m_b-1$ we have

$$T(m_b-1) = T(m_b) - \chi^{N^*-(m_b-1)} \frac{T(N^*) - T(m_b)}{\sum_{i=1}^{N^*-m_b} \chi^i}$$

or

$$T(m_b-1) = T(m_b) - \frac{(1-\chi)\,\chi^{N^*+1-m_b}}{\chi - \chi^{N^*+1-m_b}} (T(N^*)-T(m_b)) . \quad (7.4.3)$$

Since the values of $T(m_b)$, $T(m_b-1)$ and $T(N^*)$ are known, χ can be obtained from (7.4.3) as a fixed-point problem. We note that the above expression holds for $m_b \geq 2$. For $m_b=1$, we can obtain a similar fixed-point problem by setting $N=m_b$ in (7.4.1.). This necessitates the knowledge of $T(m_b+1)=T(2)$, which can be obtained numerically by solving the cyclic queueing network with 2 customers in it.

Expression (7.4.1) consistently underestimates the throughput curve. The difference between the exact throughput and the one obtained from (7.4.1) gets smaller as N gets closer to N^* or m_b, while it gets larger around the middle point $(N^*-m_b)/2$. In view of this, the following modification was introduced:

$$T(N) = T(N+1) - \alpha\chi^{N^*-N} + \alpha'_N\, x, \quad (7.4.4)$$

where

$$\alpha'_N = \begin{cases} N^*-N & N=N^*-1,\ldots,N^*-\lfloor (N^*-m_b)/2 \rfloor \\ \alpha'_{N+1} & \text{if } N=N^*-\lceil (N^*-m_b)/2 \rceil \text{ and } \lceil (N^*-m_b)/2 \rceil \neq \lfloor (N^*-m_b)/2 \rfloor \\ \alpha'_{N+1}-1 & N=m_b+1,\ldots,\lfloor (N^*-m_b)/2 \rfloor+1. \end{cases}$$

Validation tests showed that the algorithm with the above modification has a good accuracy. The average relative error in the throughput was reported to be less than 5%.

Blocking-after-service

Unlike the case of blocking-before-service, the throughput curve of a cyclic queueing network with blocking-after-service is not symmetrical with respect to N. It was conjectured that T(N) increases as N increases until it reaches a maximum value for some N* (there may be more than one value of N for which T(N) is maximum), and for N>N* T(N) is non-increasing on N. This conjecture was supported by exact numerical results. In figure 7.5 we give an example of the throughput curve of a cyclic queueing network as N changes from 1 to B=9.

From theorem 1 given in section 7.2 we have that if $m_\ell \geq B-m_\ell$, then the cyclic queueing network has the same rate matrix for all $B-m_\ell+1 \leq N \leq m_\ell$. Also, the cyclic queueing network has the same rate matrix when $N=m_\ell+1$ given that all the states in which node ℓ is blocking a node are aggregated to a single state. Since node ℓ can only block one node, this aggregate state consists of only the state in which node ℓ is blocking its upstream node. Thus, the resulting rate matrix after the aggregation takes place is in fact the rate matrix of the cyclic queueing network when $N=m_\ell+1$. In view of this, the cyclic queueing network has the same rate

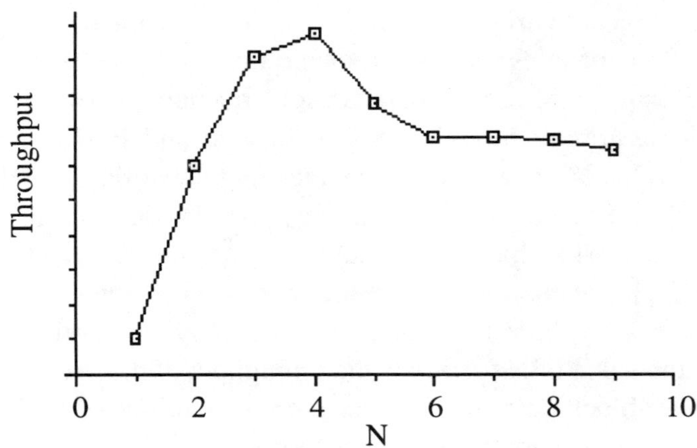

Figure 7.5: Throughput vs N (blocking-after-service).

matrix for the range of values $B-m_\ell+1 \leq N \leq m_\ell+1$. Therefore, for $B-m_\ell+1 \leq N \leq m_\ell+1$ the throughput of the network is the same. In this case, it is possible to eliminate the flat part corresponding to $B-m_\ell+1 \leq N \leq m_\ell+1$ by setting $m_\ell=B-m_\ell+1$. This has the effect of shifting to the left the part of the curve corresponding to $N \geq m_\ell+1$. As an example, consider a cyclic network for which B=12 and $m_\ell=8$. Then, the rate matrix will be identical for N=5,...,9, and thus T(5)=...=T(9). The throughput curve without the flat part will be equal to T(N), N=1,2,...,5, T(10), T(11) and T(12), N=6,7,8. This is the throughput curve of the original network in which node ℓ's buffer capacity is set equal to 5.

The throughput curve is approximated by fitting two separate curves. Using expression (7.4.4), a curve is fitted through points $T(m_b)$, $T(m_b+1)$ and T(N*), and another curve is fitted through points T(N*), T(B-1) and T(B). For $N=1,2,...,m_b$, the network has a product-form solution. Also, from the second result given in section 6.5, we have that a cyclic queueing network has a product-form solution when $N=m_b+1$. Therefore, the values $T(1),...,T(m_b)$, $T(m_b+1)$ can be easily calculated exactly. Now, let us consider the derivation of T(N*). As in the case of blocking-before-service, we obtain T(N*) by analyzing the network numerically for N=N*. However, in this case the calculation of N* is not an easy task. In general, N* depends not only on the buffer capacities but also on the service rates. N* is obtained approximately using the following equivalence between cyclic queueing networks under blocking-after-service and blocking-before-service, described in Chapter 1. Let QN1 be a cyclic queueing network consisting of M nodes with blocking-after-service, buffer capacities m_i, and exponentially distributed service times with rate μ_i, i=1,2,...,M. Let QN2 be a cyclic queueing network identical to QN1 but assuming blocking-before-service and buffer capacities $m_i'=m_i+1$, i=1,2,...,M. Then, these two queueing networks have the same rate matrix for the range $1 \leq N \leq \min_i\{m_i'+m_{i+1}'\}-1$. In view of this, the two queueing networks have the same throughput for N=1,2,..., $\min_i\{m_i'+m_{i+1}'\}-1$. This equivalence does not give us enough information to determine exactly the value of N*. Approximately, however, it seems to be equal to the value of N for which the throughput of the cyclic queueing network under blocking-before-service is maximum minus 1. Hence, we assume that the maximum throughput occurs at

$$N^* = \lceil \sum_{i=1}^{M} (m_i+1)/2 \rceil - 1.$$

T(B-1) can be computed numerically or it can be approximately constructed from T(B). Experimentation showed that the following expression for T(B-1) gave good results:

$$T(B-1) = T(B)+0.975[T^I(3)-T^I(2)],$$

where $T^I(N)$ is the throughput of the cyclic queueing network with N customers assuming that each node has an infinite buffer capacity. Finally, T(B) can be calculated as follows. Let us consider the cyclic network at an arbitrary time t assuming that N=B. Some of the servers will be busy serving and the remaining servers will be blocked. When a server completes its service, it becomes blocked immediately. When the last server completes its service, all servers will be blocked and we will have a deadlock. This deadlock is resolved by simply moving simultaneously each blocked customer to its destination node. All servers will then become unblocked and they will all start a new service at the same time. The mean time between two such successive instances is equal to $E[\max\{y_1,y_2,...,y_M\}]$, where y_i is the service time at node i, i=1,2,...,N. Since each y_i is distributed exponentially, we have

$$P(\max_i\{y_i\} < t) = \prod_{i=1}^{M} P(y_i < t),$$

$$= \prod_{i=1}^{M} (1 - e^{-\mu_i t}).$$

Hence,

$$E[\max\{y_1,y_2,...,y_M\}] = \int_0^{\infty} [1-\prod_{i=1}^{M} (1 - e^{-\mu_i t})]dt.$$

Validation tests showed that the algorithm has a good accuracy. The average relative error in the throughput was reported to be less than 5%. The critical step in this algorithm is the calculation of N^* for which the throughput is maximum. N^* can be computed as follows in the case of cyclic queueing networks where one node has an infinite buffer capacity. Let us assume that node 1 has an infinite buffer capacity. As it was mentioned above, we have that if $m_\ell \geq B-m_\ell$, then the cyclic queueing network has the same rate matrix for all $B-m_\ell+1 \leq N \leq m_\ell+1$. This range of values can be written as $B'+1 \leq N \leq m_\ell+1$, where B' is the sum of the buffer capacities of nodes 2 to M. Since $m_1=\infty$ we have that $m_1 \geq B-m_1$, and the throughput of the network is the same for all $N \geq B'+1$. Now, we can conjecture that the throughput curve increases as N increases until it reaches a maximum value and from there on it remains flat. This maximum value is attained at $N^*=B'+1$.

7.5. Symmetric cyclic queueing networks with blocking

In this section, we analyze a cyclic queueing network in which all the nodes are indistinguishable. That is, $m_i=m$ and $\mu_i=\mu$, $i=1,2,...,M$. We refer to such cyclic queueing networks as *symmetric*. In this case, the exact solution can be obtained numerically using a significantly reduced state space. For presentation purposes we will consider blocking-after-service, though the algorithm is applicable to any blocking mechanism.

Consider a three-node cyclic queueing network with four customers and m=2. The state of this network can be described by the vector (n_1,n_2,n_3), where n_i is the number of customers in the ith node, $n_i=0,1,2,3$. The value $n_i=3$ (=m+1) indicates that the ith node is full and it is blocking the previous node. Let $p(n_1,n_2,n_3)$ be the probability that the network is in state (n_1,n_2,n_3). Seeing that the nodes are indistinguishable we have

$$p(2,2,0)=p(0,2,2)=p(2,0,2)$$
$$p(2,3,0)=p(0,2,3)=p(3,0,2)$$
$$p(2,1,1)=p(1,2,1)=p(1,1,2)$$
$$p(3,1,1)=p(1,3,1)=p(1,1,3)$$

We aggregate the states (n_1,n_2,n_3) into lumps so that each lump contains all the states which have the same steady-state probability. We obtain the lumps

$$\Omega_1=\{(2,2,0),\ (0,2,2),\ (2,0,2)\}$$
$$\Omega_2=\{(2,3,0),\ (0,2,3),\ (3,0,2)\}$$
$$\Omega_3=\{(2,1,1),\ (1,2,1),\ (1,1,2)\}$$

$$\Omega_4=\{(3,1,1),\ (1,3,1),\ (1,1,3)\}.$$

This aggregate matrix can be easily constructed from the original rate matrix. For example, the transition rate from Ω_1 to Ω_2 is equal to $p((2,2,0)|\Omega_1)\mu+p((0,2,2)|\Omega_1)\mu+p((2,0,2)|\Omega_1)\mu$, where $p((n_1,n_2,n_3)|\Omega_i)$ is the probability that the network is in state (n_1,n_2,n_3) given that it is in lump Ω_i. In view of the equal probabilities identified above, we have that $p(.|\Omega_1)=1/3$. Thus, the transition rate from Ω_1 to Ω_2 becomes equal to μ. Following similar arguments, we obtain the transition rates

$$\begin{array}{c} \\ \Omega 1 \\ \Omega 2 \\ \Omega 3 \\ \Omega 4 \end{array}\begin{array}{c} \begin{array}{cccc} \Omega 1 & \Omega 2 & \Omega 3 & \Omega 4 \end{array} \\ \begin{pmatrix} * & \mu & \mu & 0 \\ 0 & * & \mu & 0 \\ \mu & 0 & * & \mu \\ \mu & \mu & 0 & * \end{pmatrix} \end{array},$$

where * indicates the negative sum of all the row elements. The steady-state probabilities $p(\Omega_i)$ can now be obtained numerically using the above rate matrix. Having determined $p(\Omega_i)$, $i=1,2,3,4$, we can obtain the individual steady-state probabilities $p(n_1,n_2,n_3)$. The following algorithm summarizes this procedure.

Step 1: Generate the aggregate rate matrix.

Step 2: Solve the system numerically to obtain $p(\Omega_i)$, $i=1,2,\ldots,\omega$, where ω is the total number of lumps.

Step 3: Calculate the normalizing constant G using the expression $G=\sum_{i=1}^{\omega}\omega_i p(W_i)$, where ω_i is the number of states in lump i.

Step 4: Obtain $p(n_1,n_2,\ldots,n_M)=G^{-1}p(\Omega_i)$, for each $(n_1,n_2,\ldots,n_M)\in\Omega_i$, $i=1,2,\ldots,\omega$.

7.6. Approximate analysis of arbitrary configurations of closed queueing networks with blocking

In this section, we present three different approximation algorithms for the analysis of arbitrary configurations of queueing networks with blocking. The first algorithm is based on the notion of *exponentialization* and it was developed to analyze closed queueing networks under the blocking mechanism of repetitive service with random destination. The second algorithm is based on Norton's theorem and it was used to analyze BCMP type of queueing networks which comprise a subnetwork of finite capacity queues. Finally, the third algorithm is based on the notion of maximum entropy and it is an extension of the maximum entropy algorithm presented in section 5.3. Validation tests showed that these three algorithm have a good accuracy.

The exponentialization algorithm

Let us consider a closed queueing network under the blocking mechanism of repetitive service with random destination. Each node i is served by k_i identical servers, and it has a maximum capacity of m_i. The service time at each node is assumed to have a phase-type distribution. Let $\mu_i(n_i)$ be the output rate of node i when there are n_i customers, where $\mu_i(0)=0$ and $\mu_i(n_i)>0$, $n_i=1,2,\ldots,m_i$. The topology of the queueing network is assumed to be such that if each service time distribution is substituted by an exponential distribution with the same mean, then the resulting queueing network becomes reversible and thus it has a product-form solution, see Chapter 6. Let N and M be the total number of customers and the total number of nodes respectively in the queueing network.

The approximation algorithm is based on the notion of *exponentialization*. The following are the main steps of the algorithm.

Step 1: For each node i, i=1,2,...,M, substitute its service time by an exponential distribution with the same rate $\mu_i(n_i)$, $n_i=0,1,\ldots,m_i$. The resulting exponential network is reversible and, therefore, it has a product-form solution. Thus, the marginal queue-length distribution $p_i(n_i)$, $n_i=0,1,\ldots,m_i$, can be derived for each node i, i=1,2,...,M.

Step 2: Derive the state-dependent arrival rate $\lambda_i(n_i)$ to each node i, i=1,2,...,M, using the expression

$$\lambda_i(n_i) = \mu_i(n_i+1)p_i(n_i+1)/p_i(n_i),\ n_i=0,1,\ldots m_i.$$

Step 3: Analyze each node i in isolation with its original phase-type service time distribution and Poisson arrivals with a state-dependent arrival rate $\lambda_i(n_i)$ obtained from step 2. Derive the marginal queue-length distribution $q_i(n_i)$, n_i=0,1,...,m_i.

Step 4: Using these new queue-length distributions calculate for each node i a new state-dependent output rate $\lambda'_i(n_i)$ from the expression

$$\lambda'_i(n_i) = \lambda_i(n_i-1)q_i(n_i-1)/q_i(n_i),\ n_i=0,1,\dots,m_i.$$

Step 5: If $\max_{i,n_i}|\lambda'_i(n_i)-\mu_i(n_i)|<\varepsilon$, then stop. Else, set $\mu_i(n_i)=\lambda'_i(n_i)$, n_i=0,1,...,m_i, i=1,2,...,M, and go back to step 1.

The algorithm can be used to obtain marginal queue-length probability distributions.

An algorithm based on Norton's theorem

We consider a closed queueing network of the BCMP type in which some of the nodes have a finite capacity. These finite capacity nodes are linked together so that to form a single subnetwork. Each finite capacity node is served by a single exponential server and blocking-before-service is assumed. Due to the finiteness of these nodes and due to the fact that they are arbitrarily interconnected, blocking of different nodes may cause a deadlock. The deadlock is resolved by simultaneously exchanging the blocked customers as described in section 7.2.

The approximation algorithm is based on Norton's theorem (see Chandy, Herzog and Woo (1975)). The queueing network is first separated into two subnetworks, namely, the *blocking subnetwork* and the *non-blocking subnetwork*. The blocking subnetwork comprises all the finite capacity nodes plus all those infinite capacity nodes which are directly upstream from the finite capacity nodes, since they are liable to getting blocked. The remaining infinite capacity nodes are grouped into the non-blocking subnetwork. For example, consider the queueing network shown in figure 7.6. Nodes 1, 2 and 3 have an infinite capacity, and nodes 4 and 5 have a finite capacity. The blocking subnetwork consists of nodes 4, 5 and 3, and the non-blocking subnetwork consists of the remaining nodes 1 and 2. The approximation procedure involves the following steps.

Step 1: Analyze the non-blocking subnetwork (obtained from the original network by "shorting" the blocking subnetwork) as a

BCMP type of queueing network assuming n customers in it, n=1,2,...,N. For each n obtain the steady-state probabilities $p(\mathbf{n}|n)$, $\mathbf{n} \in S_n$, where $\mathbf{n}$ is the state of the non-blocking subnetwork and S_n is the set of all feasible states for given n. Based on these results calculate the throughput T(n), n=1,2,...,N. Figure 7.7 gives the non-blocking subnetwork with the blocking subnetwork shorted out for the example shown in figure 7.6.

Step 2: Construct a composite node with state-dependent throughput equal to T(n), n=1,2,...,N. In the original network substitute the non-blocking subnetwork by its equivalent composite node. Analyze numerically the reduced network with N customers, and calculate the marginal queue-length probability distribution for each node. Figure 7.8 gives the blocking subnetwork with the composite queue for the example shown in figure 7.6.

Step 3: Let $p_c(n)$, n=1,2,...,N, be the marginal queue-length probability that there are n customers in the composite node (calculated in Step 2). Then, from Step 1 we have $p(\mathbf{n})=p(\mathbf{n}|n)p_c(n)$, $\mathbf{n} \in S_n$, and n=1,2,...,N. Using $p(\mathbf{n})$ the marginal queue-length probability distribution for each node in the non-blocking network can be calculated as well.

This method yields the queue-length distribution at each node. Due to the fact that we analyze the blocking subnetwork numerically, it is possible to assume different types of blocking mechanisms. We note,

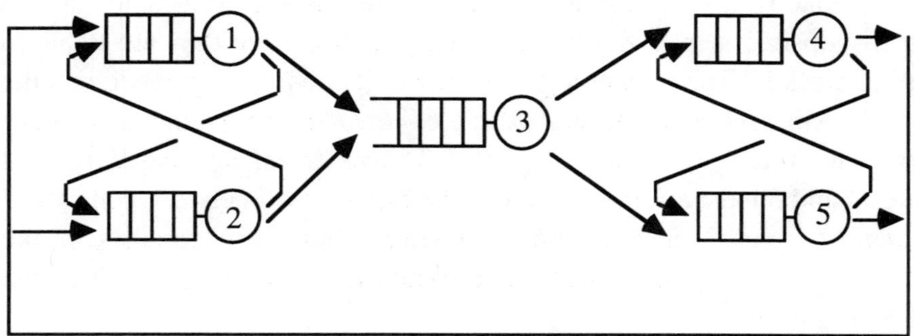

Figure 7.6: An example of the blocking and non-blocking subnetworks.

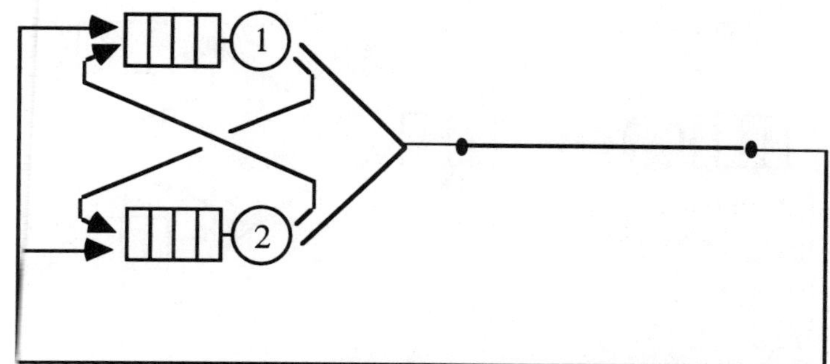

Figure 7.7: Blocking subnetwork shorted.

however, that numerical methods are not in general very efficient when analyzing large blocking subnetworks. The approximation algorithm has a good accuracy. The average relative error in the queue-length distribution of each node was reported to be less than 5%.

A maximum entropy algorithm

This algorithm is an extension of the maximum entropy algorithm presented in section 5.3 for open queueing networks with blocking. The algorithm will be presented assuming repetitive-service blocking with random destination, though it has been extended to other blocking mechanisms. Consider a closed queueing network consisting of M arbitrarily linked nodes. Each node i has a finite capacity m_i and a GE service time. Let N be the total number of customers in the network. We have that $m_i<N$, $i=1,2,...,M$. The state of the queueing network is described by the vector $\mathbf{n}=(n_1,n_2,...,n_M)$, where n_i is the number of customers in node i. Let $p(\mathbf{n})$ be the steady-state probability that the queueing network is in state $\mathbf{n}$. Let $p_i(n_i)$ be the marginal steady-state probability that there are n_i customers in the ith node, $\tilde{n}_i \le n_i \le m_i$, where $\tilde{n}_i$ is the smallest number of customers always present in node i, i.e.

$$\tilde{n}_i = \max\{0, N-\sum_{j\neq i} m_j\}.$$

As in the case of the open queueing network, the form of $p(\mathbf{n})$ can be specified by maximizing the entropy function

$$H(p) = \sum_{\mathbf{n}} p(\mathbf{n})\log p(\mathbf{n}),$$

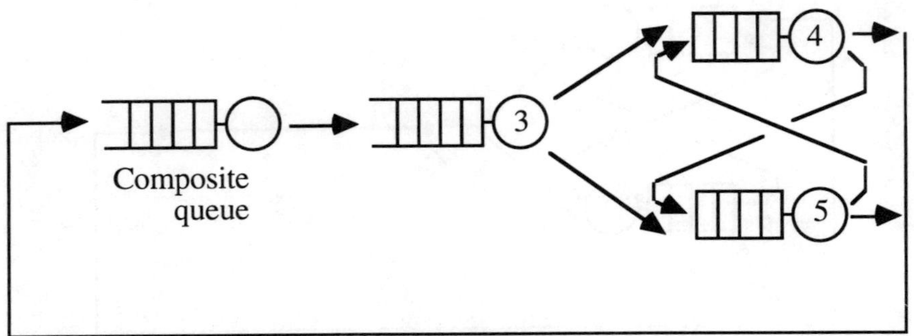

Figure 7.8: Blocking subnetwork and composite queue.

subject to the following constraints:

1. Normalization:

$$\sum_{\mathbf{n}} p(\mathbf{n}) = 1.$$

2. The probability $p_i(n_i > \tilde{n}_i)$ that there are more than $\tilde{n}_i$ customers in node i is u_i, i=1,2,...,M. That is

$$\sum_{n_i=\tilde{n}_i+1}^{m_i} p_i(n_i) = u_i.$$

3. The mean number of customers in node i is equal to L_i. That is

$$\sum_{n_i=\tilde{n}_i}^{m_i} \upsilon_i(n_i) p_i(n_i) = L_i, \text{ where } \upsilon_i(n_i)=\max\{0, n_i-(\tilde{n}_i+1)\}.$$

4. The probability that node i is full is f_i, i=1,2,...,M. That is

$$\sum_{n_i=\tilde{n}_i}^{m_i} \zeta_i(n_i) p_i(n_i) = f_i, \text{ where } \zeta_i(n_i)=\max\{0, n_i-m_i+1\}.$$

Using Lagrange's method of undetermined multipliers, we obtain

$$p(\mathbf{n}) = \frac{1}{G} \prod_{i=1}^{M} \tilde{x}_i(n_i)\, \tilde{y}_i^{\upsilon_i(n_i)}\, \tilde{z}_i^{\zeta_i(n_i)}, \tag{7.6.1}$$

where

$$\tilde{x}_i(n_i) = \begin{cases} 1 & n_i = \tilde{n}_i \\ \tilde{x}_i & \tilde{n}_i < n_i \le m_i \end{cases}, i=1,2,\ldots M,$$

G is the normalizing constant, and $\tilde{x}_i$, $\tilde{y}_i$ and $\tilde{z}_i$ are the Lagrange coefficients corresponding to constraints (2), (3) and (4). From (7.6.1) we have

$$G = \sum_{\mathbf{n}} \prod_{i=1}^{M} \tilde{x}_i(n_i)\, \tilde{y}_i^{\upsilon_i(n_i)}\, \tilde{z}_i^{\zeta_i(n_i)} . \tag{7.6.2}$$

We note that it is not possible to interchange the sum and the product in expression (7.6.2) as in the case of the open queueing network (see expression (5.3.19)). In view of this, we cannot analyze the queueing network by decomposing it into individual nodes. The closed queueing network is first analyzed as a special open queueing network in which no arrivals and departures occur. This queueing network is referred to as the *pseudo* open queueing network. From this analysis, we obtain $\tilde{x}_i$, $\tilde{y}_i$ and an initial estimate for $\tilde{z}_i$,i=1,2,...,M. Subsequently, $\tilde{z}_i$, i=1,,2,...,M, is estimated iteratively so that the traffic equations of the closed queueing network under study are satisfied. We now proceed to discuss the approximation algorithm. We will use the same notation as in section 5.3.

The pseudo open queueing network is an open queueing network with a topology identical to that of the closed queueing network under study. That is, it has the same number of nodes, server characteristics, and branching probabilities. In addition, we assume that the effective external arrival rate $\bar{\lambda}_{0i}$ is zero and that no departures occur to the outside, i.e. the branching probabilities r_{i0} are zero, i=1,2,...,M. Consequently, the traffic equations for the pseudo network are

$$\bar{\lambda}_i = \sum_{j=1}^{M} r_{ji}\, \bar{\lambda}_j, \qquad i=1,2,...,M, \tag{7.6.3}$$

where $\bar{\lambda}_i$ is the total effective arrival rate to node i (or the throughput of node i). Furthermore, we assume that the

$$N = \sum_{i=1}^{M} L_i, \tag{7.6.4}$$

where L_i is the mean number of customers in node i of the pseudo open queueing network. This pseudo open queueing network can be analyzed using the maximum entropy algorithm presented in section 5.3. In this algorithm, the traffic equations are solved first, and then the algorithm iterates round the quantities $\bar{c}^2_{di}$ (see steps 1 to 4 of the summary of the approximation algorithm). However, in the case of the pseudo network we cannot solve equations (7.6.3), since they are all dependent. Consequently, we assume that the effective arrival rate $\bar{\lambda}_k$ of some arbitrary node k is known. This enables us to solve (7.6.3) for $\bar{\lambda}_i$, $i\neq k$, $i=1,2,...,M$. The unknown $\bar{\lambda}_k$ is obtained iteratively along with the quantities $\bar{c}^2_{di}$, $i=1,2,...,M$. This is achieved by including equation (7.6.4) as an additional non-linear equation in the approximation algorithm. Using this approximation algorithm we can calculate the Lagrange coefficients x_i, y_i and z_i for each node i of the pseudo open queueing network. The Lagrange coefficients $\tilde{x}_i$ and $\tilde{y}_i$ of the closed queueing network under study are then obtained by setting $\tilde{x}_i=x_i$ and $\tilde{y}_i=y_i$, $i=1,2,...,M$. An initial estimate of $\tilde{z}_i$ is obtained by setting $\tilde{z}_i=z_i$, $i=1,2,...,M$.

Using the Lagrange coefficients $\tilde{x}_i$, $\tilde{y}_i$ and $\tilde{z}_i$, $i=1,2,...,M$, we can calculate the joint probability $p(\mathbf{n})$ from (7.6.1) and subsequently the marginal probabilities $p_i(n_i)$, $\tilde{n}_i\leq n_i\leq m_i$, $i=1,2,...,M$. The throughput T_i of node i is

$$T_i = \begin{cases} \bar{\mu}_i \sum_{n_i=\tilde{n}_i+1}^{m_i} [1-(1-r_i)^{n_i-\tilde{n}_i}]\, p_i(n_i) & ,\tilde{n}_i\geq 1 \\ \bar{\mu}_i \sum_{n_i=1}^{m_i} p_i(n_i) & ,\tilde{n}_i=0 \end{cases} \tag{7.6.5}$$

where $r_i=2/(1+\bar{c}^2_{si})$, and $\bar{\mu}_i$ and $\bar{c}^2_{si}$ is the effective service rate and the squared coefficient of variation of the effective service time respectively at node i, $i=1,2,...,M$. Now, throughputs are not the same as throughputs $\bar{\lambda}_i$, $i=1,2,...,M$, of the original closed queueing network since they were obtained from the pseudo queueing network under the assumption that

$$N = \sum_{i=1}^{M} L_i$$

rather than

$$N = \sum_{i=1}^{M} n_i.$$

However, they have to satisfy the condition

$$T_i = \sum_{j=1}^{M} r_{ji} T_j, \qquad i=1,2,...,M.$$

Now, by comparing the above condition with (7.6.3), it can be argued that

$$\frac{\bar{\lambda}_i}{T_i} = \text{constant}, \ i=1,2,...,M,$$

and therefore

$$\sum_{i=1}^{M} \frac{\bar{\lambda}_i}{T_i} L_i = \frac{\bar{\lambda}_i}{T_i} N, \ i=1,2,...,M. \qquad (7.6.6)$$

The Lagrange coefficients $\tilde{z}_i$, i=1,2,...,M, are obtained as follows.

Step 0: Set $\tilde{z}_{i0}^{(0)}=1$, i-1,2,...,M, and $\tilde{z}_i^{(0)}=\tilde{z}_i$, where $\tilde{z}_i$ was obtained from the pseudo open queueing network.

Step 1: Calculate $\tilde{z}_i = \tilde{z}_{i0}^{(0)} \tilde{z}_i^{(0)}$

Step 2: Calculate a new value of $\tilde{z}_{i0}$ from the relation

$$\tilde{z}_{i0} = \tilde{z}_{i0}^{(0)} \frac{\bar{\lambda}_i}{T_i} N \left\{ \sum_{j=1}^{M} \frac{\lambda_j}{T_j} L_i \right\}^{-1}, \ i=1,2,...,M.$$

Step 3: Check if expression (7.6.6) is satisfied. If it is satisfied then stop. Else, calculate a new solution for p(**n**), calculate the marginal probabilities $p_i(n_i)$, $\tilde{n}_i \le n_i \le m_i$, and the throughput T_i from (7.6.5), i=1,2,...,M. Set $\tilde{z}_{i0}^{(0)}=\tilde{z}_{i0}$ and go to step 1.

The approximation algorithm is more accurate than the maximum entropy approximation algorithm for open queueing networks with blocking described in section 5.3. The average relative error in the queue-length distribution and mean queue-length of each node was reported to be about 5%.

7.7. References

The theorems presented in section 7.2 are due to Onvural and Perros (1988/89). The material given in section 7.3 where it was shown that the throughput of a cyclic queueing network with blocking-before-service is symmetric with respect to the total number of customers N, is due to

Dallery and Towsley (1991). The approximate throughput analysis of cyclic queueing networks with blocking given in section 7.4 is due to Onvural and Perros (1989). Theorem 1 given in section 7.4 is due to Shanthikumar and Yao (1989). The symmetric cyclic queueing network with blocking described in section 7.5 is due to Onvural (1987) and De Nitto Personè and Grillo (1986) and (1987). The exponentialization method in section 7.6 is due to Yao and Buzacott (1986). The second approximation algorithm which is based on Norton's theorem is due to Perros, Nilsson, and Liu (1988). Finally, the maximum entropy algorithm is due to Kouvatsos and Xenios (1989a) and (1989b). The authors also developed an iterative method for calculating the normalizing constant G. An improved version of this method was proposed by Kouvatsos and Denazis (1992). Further references related to the topics presented in this Chapter are given below.

Monotonicity of the throughput of cyclic queueing networks with blocking

Shanthikumar and Yao (1989) considered cyclic queueing networks with exponentially distributed service times and blocking-before-service. They obtained conditions under which the performance measures of such networks are monotone in service rate, buffer capacity, and population size. Adan and van der Wal (1989) considered feed-forward tree-like configurations and assembly-type configurations of open queueing networks with general service times and blocking-after-service. They showed that the throughput in these networks increases if the capacity of one or more nodes is increased. See also Tsoucas and Walrand (1989).

Approximate analysis of cyclic queueing networks with blocking

Several approximation algorithms for analyzing cyclic queueing networks with blocking have been proposed. Suri and Diehl (1984) and (1986) studied approximately such queueing networks assuming that the first node has an infinite capacity. Service times were exponentially distributed and blocking-after-service or blocking-before-service was assumed. The approximation algorithm is based on the idea of replacing all the downstream nodes of the ith node, i.e. nodes i+1 to M, by a single flow-equivalent finite capacity node. This representation does not provide information as to whether the (i+1)st node is full or not. Consequently, one can not determine if the server of the ith node is blocked or not. In order to make up for this, the buffer capacity of the flow-equivalent node is allowed to vary. This flow-equivalent node is then used to capture the "view" of the server of the ith node. Dallery and Frein (1989) proposed an approximation algorithm using single-node decomposition for the analysis of cyclic queueing networks with blocking. Service times were exponentially distributed and blocking-after-service was assumed. Frein and Dallery (1989) extended this algorithm to the case of blocking-before-service. Finally, Krisht and Askin (1988) proposed an alternative approximation algorithm which involves the creation of virtual buffers to hold blocked jobs.

Approximate analysis of arbitrary configurations of closed queueing networks with blocking

Akyildiz (1988a) reported on an approximation algorithm for obtaining the throughput of closed exponential queueing networks with blocking. Blocking-after-service was assumed. The approximation algorithm is based on the idea of constructing an approximately equivalent queueing network with infinite capacity nodes which has a product-form solution. This queueing network is obtained by matching the size of its state space with that of the queueing network under study. In particular, let Z be the total number of states of the queueing network with blocking. Now consider the same queueing network assuming that each node has an infinite capacity. Clearly, this queueing network has a product-form solution. Select $\hat{N}$ the total number of customers in this queueing network, so that the total number of states $\hat{Z}$ is approximately close to Z. Then, the throughput of each node in this product-form queueing network with $\hat{N}$ customers in it, can be used as an approximation to the throughput of each node in the queueing network under study. Akyildiz (1988b) and (1989) also proposed an approximation algorithm for obtaining the mean queue-length at each node of a closed exponential queueing network with blocking and blocking-after-service. See also Akyildiz (1987). Onvural (1989) obtained results that give an insight into the decomposition of closed queueing networks. A simple throughput bound for large closed queueing networks with blocking was proposed by van Dijk (1989).

The problem of deadlock free closed queueing networks with blocking was addressed by Kundu and Akyildiz (1989). The authors obtained a necessary and sufficient condition for the capacities of the nodes so that the network is deadlock free. A buffer allocation algorithm was also developed so that no deadlocks will occur. This work was extended to the case of multiple chains in Liebeherr and Akyildiz (1991).

Bibliography

Adan, I. and van der Wal, J., Monotonicity of the throughput in single server production and assembly networks with respect to the buffer sizes, in: Perros and Altiok, (Eds.), *Proc. First International Workshop on Queueing Networks with Blocking*, (North-Holland, Amsterdam, 1989) 345-356.

Akyildiz, I. F., General closed queueing networks with blocking, in: Courtois and Latouche, (Eds.), *Performance '87* (North-Holland, Amsterdam, 1987) 283-303.

Akyildiz, I. F., On the exact and approximate throughput analysis of closed queueing networks with blocking, *IEEE Trans. Soft. Eng.* **14** (1988a) 62-71.

Akyildiz, I. F., Mean value analysis for blocking queueing networks, *IEEE Trans. Soft. Eng.* **14** (1988b) 418-429.

Akyildiz, I. F., Product form approximations for queueing networks with multiple servers and blocking, *IEEE Trans. Comput.* **38** (1989) 99-115.

Chandy, K. M., Herzog, U., and Woo, L., Parametric analysis of queueing networks, *IBM J. Res. Develop.* **19** (1975) 36-42.

Dallery, Y. and Frein, Y., A decomposition method for the approximate analysis of closed queueing networks with blocking, in: Perros and Altiok, (Eds.), *Proc. First International Workshop on Queueing Networks with Blocking*, (North-Holland, Amsterdam, 1989) 193-215.

Dallery, Y. and Towsley, D., Symmetry property of the throughput in closed tandem queueing networks with finite buffers, *Oper. Res. Lett.* **10** (1991) 541-547.

Frein, Y. and Dallery, Y., Analysis of cyclic queueing networks with finite buffers and blocking before service, *Perform. Eval.* **10** (1989) 197-210.

Kouvatsos, D. D. and Denazis, S. G., *An efficient convolution procedure for general product-form approximations*, Tech. Rept. CS-18-92, Dept. of Computing, Univ. of Bradford, 1992.

Kouvatsos, D. D. and Xenios, N. P., Maximum entropy analysis of general queueing networks with blocking, in: Perros and Altiok, (Eds.), *Proc. First International Workshop on Queueing Networks with Blocking*, (North-Holland, Amsterdam, 1989a) 281-309.

Kouvatsos, D. D. and Xenios, N. P., MEM for arbitrary queueing networks with multiple general servers and repetitive-service blocking, *Perform. Eval.* **10** (1989b) 169-195.

Krisht, A. H. and Askin, R. G., *Approximate analysis of closed exponential queueing networks with finite buffers*, Tech. Rept., Systems and Industrial Engineering Dept., University of Arizona, 1988.

Kundu, S. and Akyildiz, I. F., Deadlock free buffer allocation in closed queueing networks, *Queueing Systems J.* **4** (1989) 47-56.

Liebeherr, J. and Akyildiz, I. F., *Deadlock freedom in queueing networks with multiple routing chains and finite capacities*, Tech. Rept., College of Computing, Georgia Institute of Technology, 1991.

De Nitto Personè, V. and Grillo, D., *Impact of blocking policies in closed, exponential and symmetrical networks*, Tech. Rept. #4B0586, Networks and Services Dept., Fondazione Ugo Bordoni, Italy, 1986.

De Nitto Personè, V. and Grillo, D., Managing blocking in finite capacity symmetrical ring networks, in: Moraes, de Souza e Silva, and Soares, (Eds.), *Proc. of 3rd Int. Conf. on Data Communication Systems and their Performance*, Rio de Janeiro, 1987, 225-240.

Onvural, R. O., *Closed queueing networks with finite buffers*, Ph.D. Thesis, North Carolina State University, 1987.

Onvural, R. O., On the exact decomposition of closed queueing networks with finite buffers, in: Perros and Altiok, (Eds.), *Proc. First International Workshop on Queueing Networks with Blocking*, (North-Holland, Amsterdam, 1989) 73-83.

Onvural, R. O. and Perros, H. G., Throughput analysis of cyclic queueing networks with finite buffers, *IEEE Trans. Soft. Eng.* **15** (1989) 800-808.

Onvural, R. O. and Perros, H. G., Some equivalencies between closed queueing networks with blocking, *Perform. Eval.* **9** (1988/89) 111-118.

Perros, H. G., Nilsson, A. A. and Liu, Y.-C., Approximate analysis of product-form type queueing networks with blocking and deadlock, *Perform. Eval.* **8** (1988) 19-39.

Shanthikumar, J. G. and Yao, D. D., Monotonicity properties in cyclic queueing networks with finite buffers, in: Perros and Altiok, (Eds.), *Proc. First International Workshop on Queueing Networks with Blocking*, (North Holland, Amsterdam, 1989) 325-344.

Suri, R. and Diehl, G. W., A new building block for performance evaluation of queueing networks with finite buffers, *Proc. ACM SIGMETRICS on Measurement and Modelling of Computer Systems*, (1984) 134-142.

Suri, R. and Diehl, G. W., A variable buffer-size model and its use in analyzing closed queueing networks with blocking, *Manage. Sci.* **32** (1986) 206-224.

Tsoucas, P. and Walrand, J., Monotonicity of throughput in non-Markovian networks, *J. Appl. Probab.* **26** (1989) 134-141.

van Dijk, N. M., A simple throughput bound for large closed queueing networks with finite capacities, *Perform. Eval.* **10** (1989) 153-167.

Yao, D. D. and Buzacott, J. A., The exponentialization approach to flexible manufacturing system models with general service times, *Eur. J. Oper. Res.* **24** (1986) 410-416.

CHAPTER 8
Applications

8.1. Introduction

In this Chapter, we describe three different applications of queueing networks with blocking. In section 8.2, we present a multi-stage pull-type production/inventory system. This system is modelled as a tandem configuration of finite capacity queues. In section 8.3, we model a disk I/O system as a queueing network with blocking. In section 8.4, we describe a multi-stage interconnection network consisting of 2x2 buffered switching elements. The switching elements are interconnected so that to form a Banyan network. This multi-stage interconnection network is modelled as a discrete time queueing network with blocking. Finally, relevant references are given in section 8.5.

8.2. A multi-stage pull-type production/inventory system

Production systems can be classified into pull systems and push systems. One of the factors that distinguish these two systems is the production schedule. In a push system, it is produced as much as possible without any consideration to demand. In a pull system, on the other hand, it is produced according to demand. In this section we consider a multi-stage pull-type production/inventory system consisting of production stages with intermediate work-in progress inventory buffers. We assume that there is one type of product. After the processing at the last stage is completed the products enter a finished product warehouse. Demand for the finished products arrives at the warehouse where it is possible to have backorders. Both work-in progress and finished product levels are controlled by a continuous review (R,r) inventory control policy. Set-up is involved every time production is initiated in a production stage.

The multi-stage production/inventory system consists of M stages in series with intermediate buffer storage. The first stage is assumed to always have a sufficient number of jobs to process. Jobs in each stage are served in a FIFO manner. The Mth stage feeds into a finished product warehouse, referred to as the (M+1)st stage. Demand for finished products is assumed to arrive at the warehouse according to a compound Poisson process. That is, orders arrive at the warehouse in a Poisson fashion at the rate λ and each order requires i finished products with probability d_i. An order may not be completely satisfied. In this case, the remaining unsatisfied demand is backordered. The finished products from stage M are used to satisfy backorders first and the rest is stored in the warehouse. The inventory level in the warehouse and in the intermediate buffers is controlled by an (R,r) continuous review inventory control policy. This policy works as follows. There exists a pair of control parameters R_i, r_i, i=2,3,...,M+1. Production at stage i stops as soon as the level of inventory in the next buffer, i.e. (i+1)st buffer, reaches R_{i+1}. That is, if a departing job from stage i sees $R_{i+1}-1$ jobs in the next buffer, then stage i becomes blocked. (We can assume that stage i does not really get blocked. Rather, it takes on other tasks not directly related to this particular system.) During the time that stage i is blocked, the inventory in the (i+1)st buffer is depleted through the production at stage i+1. As soon as the inventory level in the (i+1)st buffer comes down to r_{i+1}, the reorder point, stage i becomes unblocked and starts a set-up procedure. It starts producing after the set-up is completed. The last stage of the production system, stage M, may be blocked by the warehouse. The Mth stage becomes unblocked when a total of $R_{M+1}-r_{M+1}$ finished products or more are demanded from the moment stage M becomes blocked. We note that a stage cannot hold a job during the time it is blocked. Thus, the combined maximum number of jobs that can be accommodated in stage i and in its upstream buffer i is R_i+1 when stage i is not blocked, and R_i when stage i is blocked. Let Y_i and X_i be the process completion time and set-up time at stage i, i=1,2,...,M, respectively. Y_i and X_i are assumed to have a phase-type distribution with an $(\boldsymbol{\alpha}_i,\mathbf{S}_i)$ and $(\boldsymbol{\beta}_i,\mathbf{T}_i)$ representation respectively, where $\boldsymbol{\alpha}_i=(1,0,...,0)$ and $\boldsymbol{\beta}_i=(1,0,...,0)$. Let S_{ij} and T_{ij} be the (i,j)th element of matrix $\mathbf{S}_i$ and $\mathbf{T}_i$, respectively. Then, $S_{ij},T_{ij}>0$ only if $i<j$.

This production/inventory system can be modelled as a tandem configuration of finite capacity queues, as shown in figure 8.1. A stage in the production system is represented in the queueing network by a server, and a job (or finished product) by a customer. The queue in front of server 1 is assumed to be saturated. That is, there is always at least one customer waiting to be served by server 1. A node in this queueing network represents a stage and its upstream buffer. For instance, node i represents buffer i and stage i. The maximum capacity of node i is assumed to be R_i+1. The blocking mechanism is assumed to be blocking-before-service where the position in front of the server cannot be occupied during the time that the server is blocked. This variation of the blocking-before-service mechanism was introduced in section 1.2. The blocked server becomes unblocked only when the number of customers waiting for service at the next node drops to r_{i+1}.

This queueing network was analyzed approximately by decomposing it into sub-systems as follows. For each node i, i=2,3,...,M, a sub-system is constructed consisting of node i and an upstream server which represents the upstream part of the queueing network. The server of node i is modified so that to reflect the effects of blocking due to the downstream nodes. The construction of these sub-systems and their incorporation into an overall iterative scheme is described in detail in algorithm 4 in Chapter 4. We note that algorithm 4 was presented assuming blocking-after-service and no set-ups. Also, it was assumed that a server becomes unblocked when the number of customers in the downstream node drops to the maximum capacity of the node minus 1. Algorithm 4 can be modified to allow for set-ups, a general (R,r) policy, and blocking-before-service where the server cannot be occupied when blocked.

Due to the appended warehouse to the production system, the last sub-system consists of an upstream server, representing the production system, and the warehouse. This is a special queueing model in which we

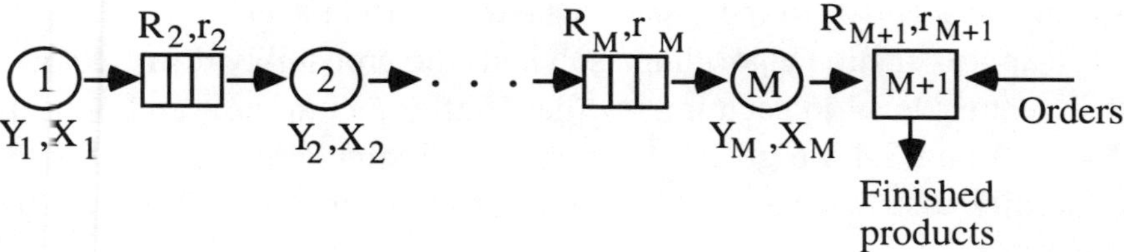

Figure 8.1: A pull-type production/inventory system. (Adopted from Altiok and Ranjan (1988))

have two types of customers. Finished products are produced by the upstream server and are stored in the warehouse. At the same time orders arrive to the warehouse, each requesting i finished products with probability d_i. When an order arrives with a request for i finished products, i finished products depart from the warehouse at the same time (bulk departure) given that i or more finished products are in the warehouse. If less than i finished products are present, say i′, then i′ finished products depart in bulk and the remaining i-i′ are backordered. The upstream server gets blocked when the inventory level in the warehouse becomes equal to R_{M+1}. It becomes unblocked at the instant when the level drops to r_{M+1}. Now, from the iterative approximation algorithm the upstream server is characterized so that, during the time the server is not blocked, we will see finished products entering the warehouse with a phase-type inter-arrival time. Also, the arrival process of the orders is assumed to be Poisson distributed. Therefore, this queueing model can be analyzed numerically.

We demonstrate the results that can be obtained and their accuracy through an example. Let us consider a production system consisting of three stages and a warehouse, i.e., M=3. Orders arrive at the warehouse at the rate of λ=0.05, and they require i finished products with probability d_i, where d_1=0.3, d_2=0.3, d_3=0.2, d_4=0.1, d_5=0.1. The (R,r) values are $(R_2,r_2)=(R_3,r_3)=(4,2)$, and $(R_4,r_4)=(8,4)$. The set-up and service times at each stage are defined by phase-type distributions whose mean and squared coefficient of variation are given in table 8.1.

In table 8.2 we give the calculated approximate values for the average inventory level in buffers 2, 3 and in the warehouse (notated by L_2, L_3, and L_4 respectively), the backorder probability, and the average backorder size. In order to get an idea of the accuracy of these approximate results, we also give estimates for the same results obtained by simulation. In general, the accuracy of the approximate results is quite good.

The approximation algorithm also yields the probability distribution of the inventory level in each intermediate buffer and in the warehouse. In tables 8.3 and 8.4 we give approximate and simulation results for the probability distribution of the inventory level in buffer 3 and in the warehouse.

	Stage 1	Stage 2	Stage 3
E(Y)	3	4	2
c_Y^2	.25	1	2
E(X)	1	2	1.5
c_X^2	2	2	2

Table 8.1: Input parameters.
(Adopted from Altiok and Ranjan (1988))

	Approximation	Simulation
L_2	2.5654	2.5238±0.0075
L_3	1.7949	1.7422±0.0147
L_4	4.9340	5.0407±0.0445
Pr(Backorder)	0.1862	0.1744
E(Backorder size)	2.1164	2.1407

Table 8.2: Approximate and simulation results.
(Adopted from Altiok and Ranjan (1988))

n_3	Approximation	Simulation
0	0.2698	0.2876
1	0.1506	0.1498
2	0.1121	0.1115
3	0.4499	0.4349
4	0.0176	0.0162

Table 8.3: Probability distribution of the inventory level in buffer 3.
(Adopted from Altiok and Ranjan (1988))

n_4	Approximation	Simulation
0	0.1165	0.1120
1	0.0395	0.0361
2	0.0512	0.0476
3	0.0677	0.0633
4	0.0801	0.0761
5	0.1583	0.1575
6	0.1452	0.1542
7	0.1149	0.1235
8	0.2257	0.2298

Table 8.4: Probability distribution of the inventory level in the warehouse.
(Adopted from Altiok and Ranjan (1988))

8.3. A disk I/O system

In this section, we describe a disk I/O configuration that supported the operations of a real-time reservation system. The disk I/O configuration shown in figure 8.2 consisted of two sets of 7 disk devices, two control units and a channel. For performance reasons, the disks were balanced. That is, the files were distributed over the 14 disks so that the rate of disk I/O requests per disk was approximately the same. Also, the mean seek time on each disk was approximately the same. The performance of the disk I/O configuration was closely monitored, and several different types of performance data were available.

The following steps are involved in the execution of a disk I/O request. The channel undertakes the transmission of a disk I/O request and its data (in the case of a write request) from the main memory to the appropriate control unit. From then on the control unit assumes full control of the execution of the disk I/O request. The channel is free to serve other disk I/O requests. The control unit initiates first a seek on the appropriate disk. This operation involves the physical movement of the read/write heads to the cylinder where the read or write request has to take place. During the time that the disk is seeking the requested cylinder the control unit is free and it can serve other disks that require attention. The disk signals to the control unit for attention upon completion of the seek. The control unit, if it is free at that moment, initiates a search on the disk. If the control unit is not free, the disk is forced to wait. At any time, there may be several disks which have completed their seek and are waiting for the control unit. These disks are served by the control unit on a FIFO basis. The search operation involves searching for the beginning of the record on the cylinder. During this operation the control unit is engaged with the particular disk and it cannot attend other disks. Upon completion of the search the data record is transmitted from the control unit to the disk in the case of a write request, or vice versa in the case of read request. In the case of the read request, there is a second transmission of the data from the control unit to the main memory. This requires the availability of the channel.

Disk I/O requests waiting execution are queued in the main memory. All disk I/O requests that are to be carried out on the same disk are placed in the same queue. Therefore, there are as many queues as the number of disks, i.e. 14. These queues have unlimited capacity and are referred to as *module queues*. Disk I/O requests in a module queue are served in a FIFO manner. The disk I/O request at the top of each module queue is automatically in service. It retains its position at the top of the queue until its service is completed.

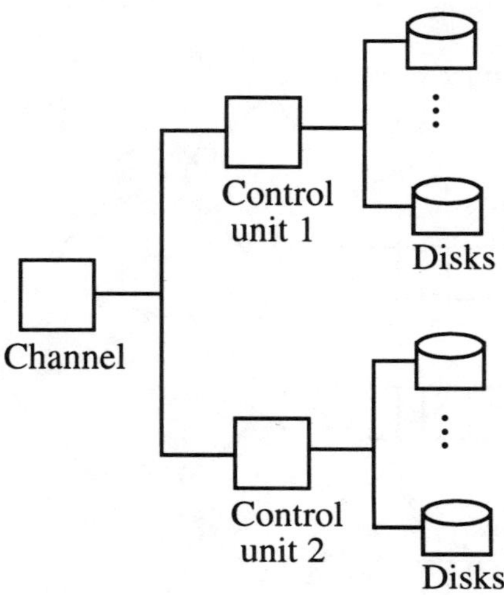

Figure 8.2: The disk I/O system.

This disk I/O configuration was modelled by the queueing network shown in figure 8.3. This queueing network consists of 14 infinite capacity queues arranged in parallel. Each queue is served by a single server, referred to as the *first-level* server. The 14 queues are grouped into two sets of seven queues. Each set of queues is linked to a single server, referred to as the *second-level* server. The two second-level servers are linked to another single server, which is referred to as the *third-level* server. All inter-arrival and service times are assumed to be exponentially distributed. Let λ, μ_1, μ_2 and μ_3 be the input rate to each queue, the service rate at a first-level server, the service rate at a second-level server, and the service rate at the third-level server respectively.

Customers in each queue are served by the corresponding first-level server on a FIFO basis. The first-level server always becomes blocked upon completion of a customer's service. The customer (call it the *blocking* customer) proceeds to the second-level server and third-level server. The first-level server remains blocked until the blocking customer departs from the queueing network. During this time no other customer in the same queue can be served. All blocking customers belonging to the same group of queues are processor-shared at the second-level server. That is, they are all served simultaneously by the same server. All blocking customers in the queueing network are processor-shared at the third-level server. A blocking customer, therefore, can be seen as cycling

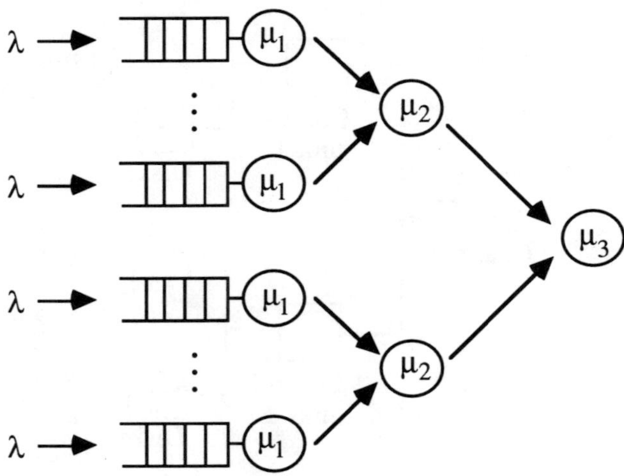

Figure 8.3: The disk I/O queueing network model.

infinitely quickly through its second-level and third-level servers, sharing the second-level server with all the blocking customers of its own group and the third-level server with all the blocking customers in the queueing network. Let $\bar{t}_b$ be the mean time a blocking customer spends in the second and third-level servers. Then, the mean time its first-level server remains blocked is also $\bar{t}_b$, and the mean effective service time at the first-level server is $(1/\mu_1)+\bar{t}_b$.

This queueing network captures the operations of the disk I/O configuration. Each of the 14 queues represents a module queue. The first-level servers represent the disks, the second-level servers represent the two control units, and the third-level server represents the channel. In this model, a disk I/O request, represented by a customer, can start its seek, i.e. its service at the first-level server, without the intervention of the channel and the control unit. This, of course, is not correct. However, the introduced error is minimal since the time to initiate a seek on a disk is negligible. The processor sharing discipline captures the heavy interaction among the disks, the control units and the channel. The assumption that the queueing network is symmetric is due to the fact that the particular disk I/O configuration was balanced. Finally, the assumption that all inter-arrival and service times are exponentially distributed was made in order to simplify the analysis of the queueing network. From performance measurements it was found that $1/\mu_1=36$ msec, $1/\mu_2=15.08$ msec, and $1/\mu_3=1.685$ msec.

We now proceed to obtain a simple approximate expression for the mean number of customers in a queue. A customer, call it the tagged customer, in a queue joins the shared servicing environment upon completion of its service. Let j_1-1 be the number of blocking customers in the group, at the time of the arrival of the tagged customer. Then, the rate of service a blocking customer will receive at the second-level server is $(1/j_1)$th the service rate that is received by a blocking customer in sole possession of the server. Thus, the tagged customer with an exponentially distributed service load t_1 will take $j_1 t_1$ of real time to complete, provided that there are j_1-1 customers constantly served with it. Similarly, let j_2-1 be the number of blocking customers sharing the third-level server at the time of the arrival of the tagged customer. Then, the tagged customer with an exponentially distributed service load t_2 will take $j_2 t_2$ of real time to complete service at the third-level server, provided that there are j_2-1 customers constantly served with it. Therefore, the total blocking time of the tagged customer is $j_1t_1+j_2t_2$. This is conditioned upon there being j_1-1 and j_2-1 blocking customers at the second and third-level servers respectively, constantly being served with the tagged customer. The unconditional mean blocking time $\bar{t}_b$ is therefore $\bar{t}_b=(J_1/\mu_2)+(J_2/\mu_3)$. Where J_1 is the expected number of blocked first-level servers (or blocking customers) in the group, conditioned upon the first-level server of the tagged customer being blocked. J_2 is the expected number of blocked first-level servers in both groups (or blocking customers) sharing the third-level server, conditioned upon the first-level server of the tagged customer being blocked. J_2 is equal to J_1 plus the unconditional expected number of blocked first-level servers in the other group. That is, $J_2=J_1+7p_b$, where p_b is the probability that a first-level server is blocked at a random instant (as opposed to serving or being idle). Now, in any queue, λ customers are processed per unit time. Therefore, the total amount of time per unit time during which a first-level server is blocked is $\lambda\bar{t}_b$, and we have $p_b=\lambda\bar{t}_b$. From the definition of J_1 we have $J_1=1+6p_b$. Therefore, we have

$$\bar{t}_b = \frac{1+6\lambda\bar{t}_b}{\mu_2} + \frac{1+13\lambda\bar{t}_b}{\mu_3}.$$

Multiplying the above expression by λ and letting $\rho_1=\lambda/\mu_1$, $\rho_2=\lambda/\mu_2$ and $\rho_3=\lambda/\mu_3$, gives

$$\lambda\bar{t}_b = (1+6\lambda\bar{t}_b)\rho_2 + (1+13\lambda\bar{t}_b)\rho_3$$

or

$$\lambda\bar{t}_b = \frac{\rho_2+\rho_3}{1-6\rho_2-13\rho_3}.$$

We now have that the traffic intensity ρ of a queue is $\rho=(\lambda/\mu_1)+\lambda\bar{t}_b$. Substituting in this expression for $\lambda\bar{t}_b$ we obtain

$$\rho = \rho_1 + \frac{\rho_2+\rho_3}{1-6\rho_2-13\rho_3}.$$

We now make the simplest assumption that the effective service time is exponentially distributed. Then the expected number L of customers in a queue is given by the formula $L=\rho/(1-\rho)$. This is in fact the expected number of disk I/O requests in a module queue. In figure 8.4, L is compared with actual data obtained from a software monitor of the disk I/O configuration. This data is in the form of observed mean number of disk I/O requests in a module queue for a given arrival rate λ, where λ varied from $3.5\text{x}10^{-3}$ to $6\text{x}10^{-3}$ disk I/O requests per msec. As can be seen this model estimates accurately the mean number of customers in a module queue. The mean waiting time W in a queue can be obtained using Little's relation, i. e. W=L/λ. This mean waiting time W is the mean disk I/O response time.

8.4. A multi-stage buffered Banyan switch

With the exception of the queueing networks with synchronized servers, this book is entirely dedicated to continuous time queueing networks with blocking. Recent advances, however, in the area of *information highways* have given rise to many interesting problems involving discrete time queueing networks with blocking. The most dominant network architecture for the information highway is the Asynchronous Transfer Mode (ATM). In ATM, each packet is exactly 53

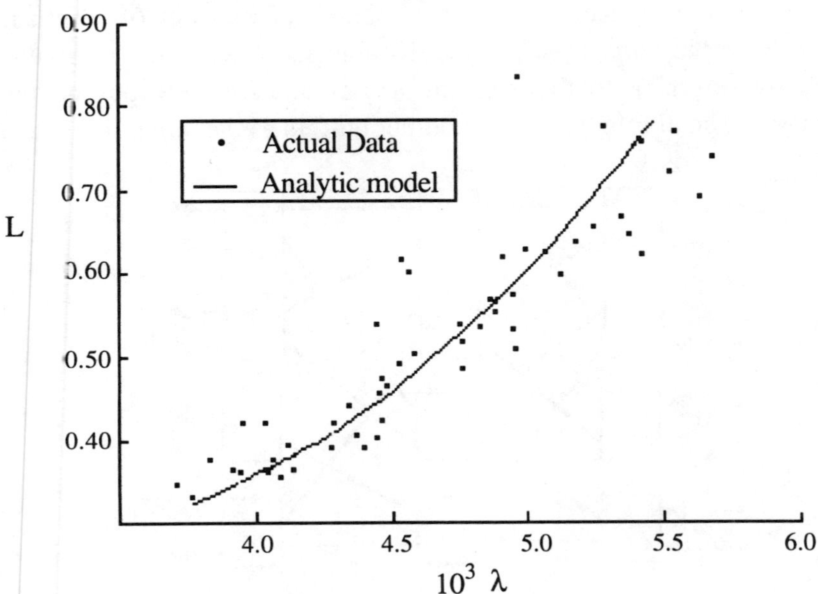

Figure 8.4: L vs actual data.

bytes long and it is referred to as a *cell*. Different ATM architectures for switching cells have been proposed so far. These switch architectures are typically modelled as discrete time queueing networks. In this last section, we present a decomposition algorithm for the analysis of a discrete time queueing network with blocking which can be used to model a buffered Banyan ATM switch architecture. As it will be seen below, the decomposition techniques developed for the analysis of continuous time queueing networks with blocking are directly applicable to this case.

Let us consider a buffered Banyan switch with switching elements which have both input and output buffering. An 8x8 Banyan switch is shown in figure 8.5, where each box represents a switching element. Each column of switching elements in figure 8.5 is known as a *stage*. Thus, in this example the switch has three stages. Stages are numbered in an ascending order from left to right. The function of the switch is to route incoming cells to its output ports. This is done by moving a cell forward from the switching elements in a stage to the switching elements in the next stage until the cell eventually departs from the switch.

Each switching element is a 2x2 crossbar as shown in figure 8.6. Each input port of the switching element has a buffer of length one, referred to as the *input buffer*. Each output port consists of a buffer of length m referred to as the *output buffer*. We distinguish between the two input buffers by referring to them as the *upper* and *lower* input buffer. Similarly, we shall refer to the two output buffers as the *upper* and *lower*

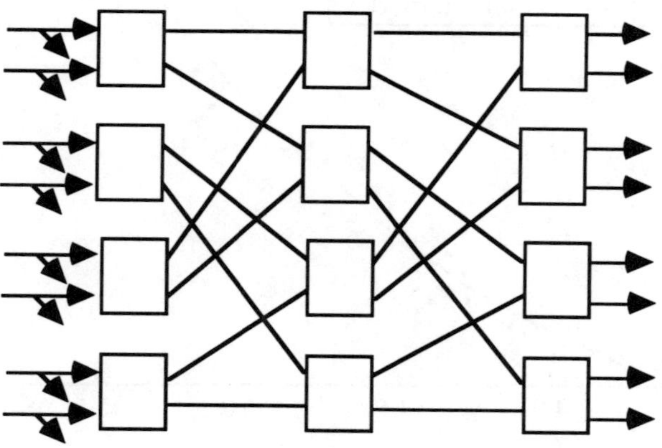

Figure 8.5: An 8X8 Banyan Network.

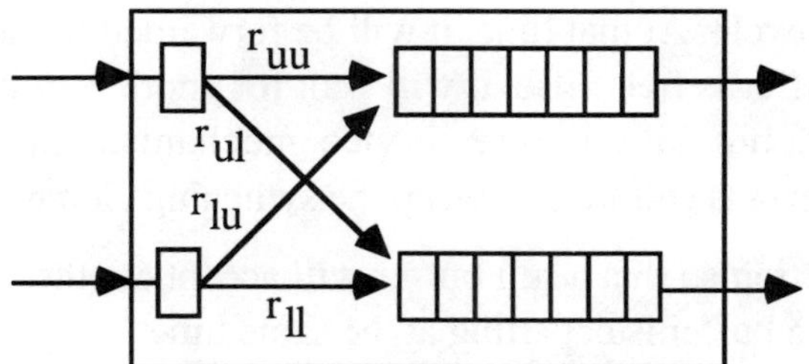

Figure 8.6: A 2x2 buffered switching element.

output buffer. Let r_{uu}, r_{ul}, r_{lu} and r_{ll} be the branching probability from the upper input buffer to the upper output buffer, from the upper input buffer to the lower output buffer, from the lower input buffer to the upper output buffer, and from the lower input buffer to the lower output buffer respectively.

The time it takes to transfer a cell from one buffer to another, within a switching element or between two adjacent switching elements, is constant. This constant time will be referred to as the *switch cycle*. All the switching elements are synchronized so that all transfers of cells from one buffer to another begin and end at the same time. Cells arrive at the switch in an asynchronous manner but they do not enter the switch until the next switch cycle. The basic operation of the switch is as follows:

1. Upon arrival to the switch, a cell waits until the next switch cycle. At that time the cell is forwarded to the corresponding input buffer if the buffer is empty. The cell is lost if the buffer is full.
2. A cell in an input buffer of a switching element is forwarded to its destination output buffer of the same switching element if there is a space. If there is no space, then the cell is blocked. That is, the cell is forced to wait in the input buffer for another switch cycle before it attempts to enter the output buffer. Blocking will also occur if both input buffers contain a cell destined to the same output buffer and the output buffer is either full or has one free space. If there is only one free space at the output buffer, then only one cell chosen randomly from the two input buffers will be forwarded.
3. A cell at the head of the output buffer is forwarded to the next switching element if the corresponding input buffer is free. The cell is blocked if the input buffer is not free. The cell will then wait until the

next switch cycle. At that time, it will be forwarded to the next stage if the input buffer is free. Else, it will wait for another switch cycle, and so on. This blocking-before-service mechanism known as *back pressure* prevents cell loss between the switching elements.

All buffers function so that a full buffer will accept another cell if the cell at the head of the buffer is departing at the same time.

The arrival process to each input port of the switch is assumed to be bursty and it is modelled by an interrupted Bernoulli process (IBP). That is, the incoming link into an input port of the switch is slotted. Each slot is assumed to be equal to the switch cycle. A slot is long enough to contain one cell. An arriving slot may or may not contain a cell. In an IBP, we have a geometrically distributed period during which no arrivals occur (idle state), followed by a geometrically distributed period during which arrivals occur in a Bernoulli fashion (active state). Given that the process is in the active state respectively idle state at slot t, it will remain in the same state in the next slot t+1 with probability p respectively q. During the active state, a slot contains a cell with probability λ. It can be shown that the average arrival rate is

$$\rho = \frac{\lambda(1-q)}{2-p-q},$$

and the squared coefficient of variation of the time between successive arrivals is

$$c^2 = 1+\lambda\left(-1+\frac{(1-p)(p+q)}{(2-p-q)^2}\right).$$

The entire switch can be viewed as a discrete time queueing network with blocking, where each node represents a switching element. The switch can be analyzed by first decomposing it into the individual switching elements, and then analyzing each switching element independently of the other switching elements. In order to analyze accurately a switching element in isolation, we need to characterize the arrival process to each input buffer of a switching element. This is done using the departure process from the output buffer of the corresponding switching element in the previous stage. Since it takes one switch cycle to forward a cell from an output buffer to the next switching element, we can think of the output buffer as being served by a server which has a constant service time equal to one switch cycle. Due to blocking, we distinguish between the *actual* and the *attempted* departure process. The actual departure process is associated with departure instants where a cell leaves the switching element. The attempted departure process involves all instants of service completion independent of whether the cell leaves or gets blocked and is forced to receive another service. In this study, we will use the attempted departure process, which we approximate by an IBP. This departure process becomes the offered arrival process to the input buffer of the corresponding switching element in the next stage. Part of this process is lost due to cells arriving to a full input buffer.

In order to analyze a switching element in isolation we also require knowledge of the blocking probability π. This is the probability that a cell at the head of an output buffer is blocked at the beginning of a switch cycle because the destination input buffer is full. This probability defines a geometric distribution which is used to represent the service at an output buffer. That is, a cell receives a service equal to a switch cycle and then with probability $1-\pi$ it departs from the switching element, or with probability π it is blocked and it is forced to receive another service.

The attempted departure process from an output buffer and the blocking probability π are calculated below. Let us assume for a moment that for a switching element we know the arrival process to each input buffer and the blocking probability π for each output buffer. Then, we can analyze the switching element numerically as follows. We first generate all the states of the system, and subsequently we generate the transition matrix **P**. The stationary probability vector **x** is then obtained by solving the system of linear equations $\mathbf{xP}=\mathbf{0}$. The state of a switching element is fully described by the variables $\mathbf{n}=\{n_0,n_1,n_2,n_3,n_4\}$. Variable n_0 gives the state of the two arrival processes to the switching element. It is 0 if both processes are in the idle state, 1 if the arrival process to the upper input buffer is in the idle state and the arrival process to the lower input buffer is in the active state, 2 if the arrival process to the upper input buffer is in the active state and the arrival process to the lower input buffer is in the idle state, and 3 if both processes are in the active state. Variable n_1 indicates the state of the upper input buffer. It is 0 if the buffer is empty, 1 if the buffer contains a new cell, 2 if the buffer contains a blocked cell destined for the upper output buffer, and 3 if the buffer contains a blocked cell destined for the lower output buffer. Variable n_2 gives the state of the lower input buffer and it takes the same values as n_1. Finally, variables n_3 and n_4 give the number of waiting cells in the upper output buffer and in the lower output buffer respectively. We have n_3, n_4=0,1,...,m, where m is the capacity of each output buffer. Non-uniform destination distributions can be easily implemented by adjusting the branching probabilities r_{uu}, r_{ul}, r_{lu} and r_{ll} within each switching element so that cells from an input port of the switch can follow a path to a specific output port of the switch with a given probability.

In order to characterize the attempted departure process from an output buffer by an IBP we have to obtain values for p, q and λ. Conceptually, the values for p and q can be obtained as follows. Let us consider the set of all states of the switching element from which it is possible to have an attempted departure from the particular output buffer. If the switching element is in one of these states, then the departure process is in its active state. We shall refer to these states as the *active states*, and their set will be referred to by the symbol A. Probability p can then be obtained as the probability that the system will be in one of the active states in the next slot, given that it is currently in an active state. Probability q can be obtained similarly by considering the set of all states of the switching element from which it is not possible to have an attempted departure from the particular output buffer. If the switching element is in one of these states, then the departure process is in its idle state. We shall refer to these states as the *idle states*, and their set will be referred to by the symbol I. In view of this, the characterization of the departure process from a particular output buffer involves the classification of the states of the switching element into active and idle states. However, this is not a straight forward task. Below, we present three different models for characterizing the departure process as an IBP.

Model 1

In model 1 we simplify the problem of characterizing the departure process by setting λ=1. Any state of the switching element from which it is possible to have an attempted departure from the output buffer under consideration is an active state, and all the others are idle states. Let us consider the attempted departure process from the upper output buffer. Then, the set of active states A consists of any state $\{n_0,n_1,n_2,n_3,n_4\}$ where $n_3>0$. Also, when $n_3=0$ states $\{n_0,1,0,0,n_4\}$, $\{n_0,1,1,0,n_4\}$ and

$\{n_0,0,1,0,n_4\}$ are active states with probability r_{uu}, $1\text{-}r_{ul}r_{ll}$ and r_{lu} respectively. The probabilities p and q can be calculated as follows. Let $p(\mathbf{n})$ be the steady-state probability that the switching element is in state **n**, and let $q(\mathbf{n},\mathbf{n}')$ be the transition probability from state **n** to state **n′**. Then, we have

$$p = \sum_{\mathbf{n}\in A} p(\mathbf{n})\Big[\sum_{\mathbf{n}'\in A} q(\mathbf{n},\mathbf{n}')\Big] \Big/ \sum_{\mathbf{n}\in A} p(\mathbf{n})$$

$$q = \sum_{\mathbf{n}\in I} p(\mathbf{n})\Big[\sum_{\mathbf{n}'\in I} q(\mathbf{n},\mathbf{n}')\Big] \Big/ \sum_{\mathbf{n}\in I} p(\mathbf{n}).$$

Model 1 becomes inaccurate when the input traffic to the switch is bursty. Typically, bursty traffic slowly loses its burstiness as it proceeds through the switch. (In this study, the burstiness of an arrival (departure) process is measured by the squared coefficient of variation c^2 of the inter-arrival (inter-departure) times.) Comparing the results from model 1 with simulation data, it was observed that the c^2 of the attempted departure process drops off to a very small value immediately after the first stage. This value is significantly lower than the c^2 obtained by simulation. As a result, the cell loss probability at the input of the switch is poorly estimated.

The blocking probability π can be directly obtained from the steady-state probabilities $p(\mathbf{n})$. Let us consider the upper input buffer. Let A_u be the set of states of the switching element in which the source to the upper input buffer is in the active state, i.e. $n_0 = 2$ or 3. Furthermore, let B be the set of all states in which the particular input buffer is blocked, i.e. $n_1=2$ or 3. If $\mathbf{n}\in A_u$, then the probability of an attempted arrival to the upper input buffer during the next slot is $\lambda_u p(\mathbf{n})$, where λ_u is the IBP arrival parameter associated with the upper input buffer. The probability that the next slot produces a blocked state is $p(\mathbf{n})q(\mathbf{n},\mathbf{n}')$, where $\mathbf{n}'\in B$. We then have

$$\pi = \sum_{\mathbf{n}\in A_u} \lambda_u p(\mathbf{n})\Big[\sum_{\mathbf{n}'\in B} q(\mathbf{n},\mathbf{n}')\Big] \Big/ \sum_{\mathbf{n}\in A_u} \lambda_u p(\mathbf{n}).$$

Model 2

The idle states are defined to be the states in which the output buffer is empty, both inputs buffers are empty, and both arrival processes to the switching element are in the idle state. The active states are all the other states. Since it is not possible to have an attempted departure from certain active states, we have that $\lambda < 1$. The length of the time that the departure process is in its active state is longer than in model 1. This is because the set of active states includes all the active states as defined in model 1, plus any state in which at least one of the arrival processes is in the active state. That is, for the departure process from the upper output buffer the active states are: $\{n_0,n_1,n_2,n_3,n_4\}$, where $n_3>0$, $\{n_0,1,0,0,n_4\}$, $\{n_0,1,1,0,n_4\}$, $\{n_0,0,1,0,n_4\}$ with probability r_{uu}, $1\text{-}r_{ul}r_{ll}$ and r_{lu} respectively, and $\{1,0,0,0,n_4\}$, $\{2,0,0,0,n_4\}$, $\{3,0,0,0,n_4\}$ with probability r_{lu}, r_{uu} and $1\text{-}r_{ul}r_{ll}$ respectively. Let p, q and λ be the parameters of the attempted departure process from an output buffer, say the upper one. Also, let π be the blocking probability associated with this output buffer. Then, p, q and π can be obtained in the same way as in model 1. Since there is no cell loss within the switch, λ can be obtained by equating throughputs at the output and input of the switching element. Let p_u, q_u, λ_u and p_l, q_l, λ_l be the parameters of the arrival process to the upper and lower input buffer respectively. These arrival processes are the attempted departure processes from the corresponding output buffers in the previous stage. Let π^u and π^l be the blocking probabilities associated with these two (previous stage) output buffers. We have

$$\lambda = \frac{2-p-q}{(1-\pi)(1-q)}\left\{\frac{r_{uu}\lambda_u(1-\pi^u)(1-q_u)}{2-p_u-q_u} + \frac{r_{lu}\lambda_l(1-\pi^l)(1-q_l)}{2-p_l-q_l}\right\}.$$

Using this new definition of idle and active states we obtain a model which accurately estimates the squared coefficient of variation c^2 of the time between two successive attempted departures. It was observed, however, that the value of λ was low when compared to simulation results. In model 3, we obtain an IBP where λ has a value higher than in model 2.

Model 3

In this model we start with the parameters obtained from model 2. That is, the definition of idle and active states and the calculation of p, q, λ and π are the same as in model 2. From p, q and λ we can obtain the average arrival rate ρ and the squared coefficient of variation c^2. Given ρ and c^2

we specify a new IBP with parameters p', q' and λ', where λ' is set equal to one, and p' and q' are obtained using ρ and c^2. We have

$$p' = 1 - \frac{2}{c^2(\tau+1)^2+\tau+1}$$

and

$$q'=1-\tau p',$$

where $\tau=\lambda/(1-\lambda)$. Note that model 3 estimates c^2 in the same way as model 2. Setting $\lambda'=1$ is an arbitrary choice. It has been observed that model 3 provides in most cases an upper bound for the performance measures of interest. In reality, a value for the peak rate between the values obtained by models 2 and 3 seem to be the best choice. Unfortunately, there does not appear to be any easy way of obtaining this value, since it depends on the traffic parameters and the topology of the switching element.

Summary of the approximation algorithm

The switch is analyzed by decomposing it into the individual switching elements. Each switching element is then analyzed individually assuming that the arrival process to each input buffer and the blocking probabilities are known. The entire switch is analyzed using the following iterative scheme. An iteration consists of visiting each switching element beginning at the first stage of the switch and continuing sequentially to the last stage of the switch. For the analysis of each switching element we use the blocking probabilities from the previous iteration and the attempted arrival processes from the current iteration.

Step 0: For each switching element assume initial values for the blocking probability for each input buffer. Set i to the first stage of the switch (i=1), i. e. i=1.

Step 1: Analyze numerically each switching element in stage i.

Step 2: For each switching element in stage i calculate the new attempted departure process from each output buffer and the new blocking probability for each input buffer using any of the three models described above. Note that the new attempted departure process becomes the input process to the next stage.

Step 3: If stage i is not the last stage of the switch, set i=i+1 and go to step 1. Else, this iteration is completed. Go to step 4.

Step 4: The convergence test compares p(**n**) for each switching element from one iteration to the next. Stop, if convergence has occurred. Else set i=1 and go to step 1.

The approximation algorithm gives the queue-length distribution for each buffer of each switching element. From this, performance measures such as cell loss probability, mean time to traverse the switch, and throughput can be obtained.

In figure 8.7 we give a numerical example for a 16x16 switch. This example is for the *hotspot pattern*, a traffic pattern that gives rise to severe blocking and congestion across the switch. This type of traffic was implemented as follows. All the arrival processes to the input ports of the switch were assumed to be identical and equal to the same IBP. However, a percentage of each input source is destined to the same output, referred to as the *hot output*. The remainder of the traffic from each input source is distributed uniformly amongst all destinations including the hot output. The output which receives the least amount of traffic is referred to as the *non-hot output*. Let ρ and c^2 be the average arrival rate and the squared coefficient of variation of the inter-arrival time of the common IBP that

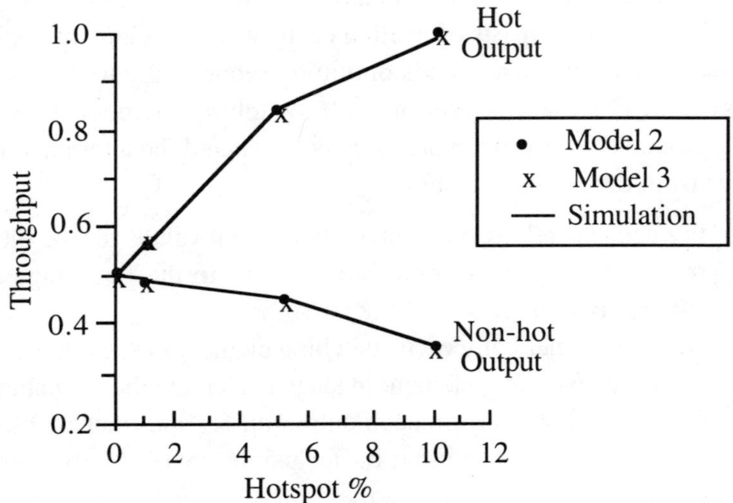

Figure 8.7: Throughput vs Hotspot %; c^2=200, ρ=0.5.

represents all input sources to the switch. For this IBP we assume that λ=1. The capacity m of each output buffer was set to 4. Figure 8.7 gives the throughput as a function of the hotspot percentage, i. e. the percentage of each input source destined to the hot output. Both models 2 and 3 were found to have good accuracy even when the throughput is close to one.

In general, the approximation algorithm has a good accuracy. The relative error of the throughput estimation was found to be less than 3%, even when the input traffic was highly bursty, i. e. c^2=200. The relative error for the mean delay estimation was between 5% and 10%. Finally, the estimation of the cell loss probability was within the same order of magnitude as the simulated results. This is acceptable since this probability is associated with very small values.

8.5. References

The multi-stage pull-type production/inventory system model and its analysis is due to Altiok and Ranjan (1988). The warehouse queueing model was analyzed by the authors as an M/G/1 queue using a method due to Altiok (1987). The disk I/O queueing network model is due to Foster and Perros (1979). An alternative disk I/O queueing network with blocking model can be found in Foster and Perros (1980). Also, a simple non-linear regression model for predicting the response time of a disk I/O system was developed by Perros (1980). The analysis of the buffered Banyan ATM switch as a discrete time queueing network with blocking is due to Morris and Perros (1992)

Bibliography

Altiok, T., Queue with group arrivals and exhaustive service discipline, *Queueing Systems* **2** (1987) 307-320.

Altiok, T. and Ranjan, R., *Multi-stage, pull-type production/inventory systems with set-ups and (R,r) control policies*, Tech. Rept. 88-108, Industrial Engineering Dept., Rutgers Univ., 1988. To appear in *IIE Trans.*

Morris, T. D. and Perros, H. G., Performance modelling of a multi-buffered Banyan switch under bursty traffic, *INFOCOM '92*, Florence, Italy, May 1992, 436-445.

Foster, F. G. and Perros, H. G., Hierarchical queue networks with partially shared servicing, *J. Opl. Res. Soc.* **30** (1979) 157-166.

Foster, F. G. and Perros, H. G., On the blocking process in queue networks, *Eur. J. Oper. Res.* **5** (1980) 276-283.

Perros, H. G., A regression model for predicting the response time of a disc I/O system, *The Computer Journal* **23** (1980) 34-36.

Bibliography

Adan, I. and van der Wal, J., Monotonicity of the throughput in single server production and assembly networks with respect to the buffer sizes, in: Perros and Altiok, (Eds.), *Proc. First Intl. Workshop on Queueing Networks with Blocking*, (North-Holland, Amsterdam, 1989) 345-356.

Akyildiz, I. F., Exact product form solution for queueing networks with blocking, *IEEE Trans. Comput.* **36** (1987) 121-126.

Akyildiz, I. F., General closed queueing networks with blocking, in: Courtois and Latouche, (Eds.), *Performance '87* (North-Holland, Amsterdam, 1987) 283-303.

Akyildiz, I. F., Mean value analysis for blocking queueing networks, *IEEE Trans. Soft. Eng.* **14** (1988) 418-429.

Akyildiz, I. F., On the exact and approximate throughput analysis of closed queueing networks with blocking, *IEEE Trans. Soft. Eng.* **14** (1988) 62-71.

Akyildiz, I. F., Performance analysis of computer and communication networks with local and global window flow control, *INFOCOM 88*, 400-410.

Akyildiz, I. F., Exact analysis of queueing networks with rejection blocking, in: Perros and Altiok, (Eds.), *Proc. First International Workshop on Queueing Networks with Blocking*, (North-Holland, Amsterdam, 1989) 19-29.

Akyildiz, I. F., Product form approximations for queueing networks with multiple servers and blocking, *IEEE Trans. Comput.* **38** (1989) 99-115.

Akyildiz, I. F. and von Brand, H., Exact solutions for open, closed and mixed queueing networks with rejection blocking, *Theoret. Comput. Sci.* **64** (1989) 203-219.

Akyildiz, I. F. and von Brand, H., Computational algorithms for networks of queues with rejection blocking, *Acta Inf.* **26** (1989) 559-576.

Akyildiz, I. F. and von Brand, H., Central server models with multiple job classes, state dependent routing, and rejection blocking, *IEEE Trans. Soft. Eng.* **15** (1989) 1305-1312.

Akyildiz, I. F. and von Brand, H., Dual and self dual networks of queues with rejection blocking, *Computing J.* **43** (1989) 1-11.

Akyildiz, I. F. and Huang, C.-C., Exact analysis of multi-job class networks of queues with blocking-after-service, in: Onvural and Akyildiz, (Eds.), *Proc. Second International Workshop on Queueing Networks with Finite Capacity*, (North-Holland, Amsterdam, 1993) 143-156.

Akyildiz, I. F. and Liebeherr, J., Application of Norton's theorem on queueing networks with finite capacities, *INFOCOM 89* 914-923.

Akyildiz, I. F. and van Dijk, N. M., Exact solution of parallel queues with finite buffers, in: King, Mitrani, and Pooley, (Eds.), *PERFORMANCE '90* (North-Holland, Amsterdam, 1990) 35-49.

Altiok, T., Approximate analysis of exponential tandem queues with blocking, *Eur. J. Oper. Res.* **11** (1982) 390-398.

Altiok, T., Production lines with phase type operations and repair times and finite buffers, *Int. J. Prod. Res.* **28** (1985) 489-498.

Altiok, T., Approximate analysis of queues in series with phase-type service times and blocking, *Oper. Res.* **37** (1989) 601-610.

Altiok, T. and Kao, Z., Bounds for throughput in production/inventory systems in series with deterministic processing times, *IIE Trans.* **21** (1989) 82-85.

Altiok, T. and Perros, H. G., Open networks of queues with blocking: split and merge configurations, *IIE Trans.* (1986) 251-261.

Altiok, T. and Perros, H. G., Approximate analysis of arbitrary configurations of open queueing networks with blocking, *Ann. Oper. Res.* **9** (1987) 481-509.

Altiok, T. and Ranjan, R., Analysis of production lines with general service times and finite buffers: A two-node decomposition approach, *Engineering Costs and Production Economics* **17** (1988) 155-165.

Altiok, T. and Ranjan, R., *Multi-stage, pull-type production/inventory systems with set-ups and (R,r) control policies*, Tech. Rept. 88-108, Industrial Engineering Dept., Rutgers Univ., 1988. To appear in *IIE Trans.*

Altiok, T. and Stidham, S., A note on transfer lines with unreliable machines, random processing times and finite buffers, *IIE Trans.* (1982) 125-127.

Altiok, T. and Stidham, S., The allocation of interstage buffer capacities in production lines, *IIE Trans.* **15** (1983) 292-299.

Altiok, T. and Shiue, G. A., *Single-stage, multi-product production/inventory systems with lost sales*, Tech. Rept. 90-101, Industrial Engineering Dept., Rutgers Univ., 1990. To appear in *IIE Trans.*

Alvarez, R., Dallery, Y., and David, R., An experimental study of the continuous flow model of transfer lines with unreliable machines and finite buffers, *Proc. of IMACS Symposium on Modelling and Control of Technological Systems*, Villeneuve d'Ascq, 1991.

Ammar, M. H. and Gershwin, S. B., Equivalence relations in queueing models of fork/join networks with blocking, *Perf. Eval.* **10** (1989) 233-245.

Artamonov, G. T., Productivity of a two-instrument discrete processing line in the presence of failures, *Kibernetica* **3** (1976) 126-130 (English translation: *Cybernetics* **12** (1977) 464-468).

Asare, B., *Queue networks with blocking*, Ph.D. Thesis, Trinity College Dublin, Ireland, 1978.

Atmaca, T., *Approximate analysis of a conveyor system*, Tech. Rept., Laboratoire MASI, Paris 6 University, 1990.

Avi-Itzhak, B., A sequence of service stations with arbitrary input and regular service times, *Manage. Sci.* **11** (1965) 565-571.

Avi-Itzhak, B. and Levy, H., *A sequence of servers with arbitrary input and regular service times revisited*, Tech. Rept. 5-93, Rutcor Research Center, Rutgers Univ., 1993.

Avi-Itzhak, B. and Yadin, M., A sequence of two servers with no intermediate queue, *Manage. Sci.* **11** (1965) 565-571.

Balsamo, S. and Clò, M. C., *State distribution at arrival times for closed queueing networks with blocking*, Tech. Rept. TR-35/92, Dipartimento di Informatica, Universita di Pisa, 1992.

Balsamo, S., Clò, M. C., and Donatiello, L., Cycle time distribution of cyclic networks with blocking, in: Onvural and Akyildiz, (Eds.), *Proc. Second International Workshop on Queueing Networks with Finite Capacity*, (North-Holland, Amsterdam, 1993) 27-40.

Balsamo, S. and Donatiello, L., On the cycle time distribution in a two-stage cyclic network with blocking, *IEEE Trans. Soft. Eng.* **15** (1989) 1206-1216.

Balsamo, S. and Iazeolla, G., Some equivalence properties for queueing networks with and without blocking, in: Agrawala and Tripathi, (Eds.), *Performance '83* , (North-Holland, Amsterdam, 1983), 351-360.

Balsamo, S. and De Nitto Personè, V., Closed queueing networks with finite capacities: blocking types, product-form solution and performance indices, *Perform. Eval.* **12** (1991) 85-102.

Balsamo, S. and De Nitto Personè, V., *A survey of product-form queueing networks with blocking and their equivalences*, Tech. Rept., Dipartmento di Ingegneria Elettronica, Universita di Roma, 1992. To appear in *Ann. Oper. Res.*

Balsamo, S., De Nitto Personè, V., and Iazeolla, G., Identity and reducibility properties of some blocking and non-blocking mechanisms in congested networks, in: Odoni, Bianco, and Szego, (Eds.), *Flow control in congested networks*, NATO ASI Series, Comp. and System Sci. Vol F38 (Springer-Verlag, 1987).

Bell, P. C., The use of decomposition techniques for the analysis of open restricted queueing networks, *Oper. Res. Lett.* **1** (1982) 230-235.

Bocharov, P. P., On the two-node queueing networks with finite capacity, in: Perros and Altiok, (Eds.), *Proc. First International Workshop on Queueing Networks with Blocking*, (North-Holland, Amsterdam,1989) 105-125.

Bocharov, P. P. and Albores, F. K., On two-stage exponential queueing system with internal losses or blocking, *Prob. Control Infor. Theory* **9** (1980) 365-379.

Bocharov, P. P. and Rokhas, G. P., On an exponential queueing system in series with blocking, *Prob. Control Infor. Theory* **9** (1980) 441-455.

Bocharov, P. P. and Spesivov, S. S., On multistage queueing system with losses and arriving flow of phase type, *Prob. Control Infor Theory* **15** (1986) 465-474.

Bouhchouch, A., Frein, Y., and Dallery, Y., A decomposition method for the analysis of tandem queueing networks with blocking before service, in: Onvural and Akyildiz, (Eds.), *Proc. Second International Workshop on Queueing Networks with Finite Capacity*, (North-Holland, Amsterdam, 1993) 97-112.

Boxma, O. and Konheim, A., Approximate analysis of exponential queueing system with blocking, *Acta Inf.* **15** (1981) 19-66.

Brandwajn, A. and Jow, Y.-L. L., An approximation method for tandem queues with blocking, *Oper. Res.* **36** (1988) 73-83.

Brandwajn, A. and Jow, Y.-L. L., Tandem exponential queues with finite buffers, in: *Proc. Int. Seminar on Computer Networking and Performance Evaluation*, Tokyo, Sept 18-20, 1985.

Brandwajn, A. and Sahai, A. K., A note on iterative solution of open tandem networks with blocking, *Perf. Eval.* **10** (1989) 211-218.

Buzacott, J. A., Automatic transfer lines with buffer stocks, *Int. J. Prod. Res.* **5** (1967) 183-200.

Buzacott, J. A., The effect of station breakdowns and random processing times on the capacity of flow lines with in-process storage, *AIIE Trans.* **4** (1972) 308-312.

Buzacott, J. A. and Hanifin, L. E., Models of automatic transfer lines with inventory banks - a review and comparison, *AIIE Trans.* **10** (1978) 197-207.

Caseau, P. and Pujolle, G., Throughput capacity of a sequence of queues with blocking due to finite waiting room, *IEEE Trans. Soft. Eng.* **5** (1979) 631-642.

Chakravarthy, S., Analysis of production line systems with two unreliable machines with phase-type processing times and a finite storage buffer, *Stochastic Models* **3** (1987) 369-391.

Chakravarthy, S., A finite capacity queueing network with single and multiple processing nodes, in: Onvural and Akyildiz, (Eds.), *Proc. Second International Workshop on Queueing Networks with Finite Capacity*, (North-Holland, Amsterdam, 1993) 197-211.

Chao, X. and Chen, H., *Optimal intensity control for tandem queues with blocking*, Tech. Rept., Dept. of Mechanical and Industrial Engineering, New Jersey Institute of Technology, 1990.

Cheng, D. W., Analysis of a tandem queue with state-dependent general blocking: A GSMP perspective, in: Onvural and Akyildiz, (Eds.), *Proc. Second International Workshop on Queueing Networks with Finite Capacity*, (North-Holland, Amsterdam, 1993) 185-195.

Choo, Q. H. and Conolly, B., Waiting time analysis for a tandem queue with correlated service, *Eur. J. Oper. Res.* **4** (1980) 337-345.

Choong, Y. F. and Gershwin, S. B., A decomposition method for the approximate evaluation of capacitated transfer lines with unreliable machines and random processing times, *IIE Trans.* **19** (1985) 150-159.

Choukri, T., Exact analysis of multiple job classes and different types of blocking, in: Onvural and Akyildiz, (Eds.), *Proc. Second International Workshop on Queueing Networks with Finite Capacity*, (North-Holland, Amsterdam, 1993) 15-26.

Clarke, A. B., *Markovian queues with servers in tandem*, Tech. Rept. 49, Mathematics Dept., Western Michigan Univ., 1977.

Clarke, A. B., *A two-server tandem queueing system with storage between servers*, Tech. Rept. 50, Mathematics Dept., Western Michigan Univ., 1977.

Clarke, A. B., *A multiserver general service time queue with servers in series*, Tech. Rept. 51, Mathematics Dept., Western Michigan Univ., 1978.

Clarke, A. B., *Waiting times for Markovian queue with servers in series*, Tech. Rept. 52, Mathematics Dept., Western Michigan Univ., 1978.

Choi, M., *Multiclass queueing networks with finite buffers*, Ph.D. Thesis, University of North Carolina State University, 1989.

Cohen, J. W., The multiple phase service network with generalized processor sharing, *Acta Inf.* **12** (1979) 245-284.

Commault, C. and Dallery, Y., Production rate of transfer lines without buffer storage, *IIE Trans.* **22** (1990) 315-329.

Conway, A. E. and Keilson, J., *Delay distribution in a two-stage finite buffer queueing model with blocking and flow control*, Tech. Rept., GTE Labs, 1992.

Courtois, P.-J. and Semal, P., Computable bounds for conditional steady-state probabilities in large Markov chains and queueing models, *IEEE JSAC* **4,6** (1986) 920-936.

Dallery, Y., David, R., and Xie, X.-L., An efficient algorithm for analysis of transfer lines with unreliable machines and finite buffers, *IEE* **20** (1988) 280-283.

Dallery, Y., David, R., and Xie, X.-L., Approximate analysis of transfer lines with unreliable machines and finite buffers, *IEEE Trans. Aut. Control* **34** (1989) 943-953.

Dallery, Y. and Frein, Y., A decomposition method for the approximate analysis of closed queueing networks with blocking, in: Perros and Altiok, (Eds.), *Proc. First International Workshop on Queueing Networks with Blocking*, (North-Holland, Amsterdam, 1989) 193-215.

Dallery, Y. and Frein, Y., On decomposition methods for tandem queueing networks with blocking, *Oper. Res.* **41** (1993) 386-399.

Dallery, Y. and Gershwin, S. B., Manufacturing flow line systems: A review of models and analytical results, *Queueing Systems J.* **12** (1992) 3-94.

Dallery, Y., Z. Liu, and Towsley, D., *Equivalence, reversibility and symmetry properties in fork/join queueing networks with blocking*, Tech. Rept., Laboratoire MASI, Paris 6 University, 1992.

Dallery, Y., Z. Liu, and Towsley, D., *Properties of fork/join queueing networks with blocking under various operating mechanisms*, Tech. Rept., Laboratoire MASI, Paris 6 University, 1993.

Dallery, Y. and Towsley, D., Symmetry property of the throughput in closed tandem queueing networks with finite buffers, *Oper. Res. Letters* **10** (1991) 541-547.

Dallery, Y. and Yao, D. D., Modelling of flexible manufacturing cells, in: Kusiak (Ed.) *Modelling and Design of Flexible Manufacturing Systems,* (North-Holland, 1986) 289-300.

Daskalaki, S. and MacGregor Smith, J., The static routing problem in open finite queueing networks, in: Perros and Altiok, (Eds.), *Proc First Int. Workshop on Queueing Networks with blocking*, (North-Holland, Amsterdam, 1989) 313-324.

Dattatreya, E.S., *Tandem queueing systems with blocking*, Ph.D. Thesis, University of California, Berkeley, 1978.

David, R., Xie, X., and Dallery, Y., Properties of continuous models of transfer lines with unreliable machines and finite buffers, *IMA J. Mathematics Applied in Business & Industry* **6** (1990) 281-308.

Decreusefond, L., Korezlioglu, H., and van Dijk, N., An error bound for infinite approximations of queueing networks with large finite stations, in: Onvural and Akyildiz, (Eds.), *Proc. Second International Workshop on Queueing Networks with Finite Capacity*, (North-Holland, Amsterdam, 1993) 239-251.

Evans, R. V., Capacity of queueing networks, *Oper. Res.* **15** (1967) 530-536.

Evans, R. V., Geometric distributions in some two-dimensional queueing systems, *Oper. Res.* **15** (1967) 830-846.

Foster, F. G. and Perros, H. G., Hierarchical queue networks with partially shared servicing, *J. Opl. Res. Soc.* **30** (1979) 157-166.

Foster, F. G. and Perros, H. G., On the blocking process in queue networks, *Eur. J. Oper. Res.* **5** (1980) 276-283.

Frein, Y. and Dallery, Y., Analysis of cyclic queueing networks with finite buffers and blocking before service, *Perform. Eval.* **10** (1989) 197-210.

Gershwin, S. B., An efficient decomposition method for the approximate evaluation of tandem queues with finite storage space and blocking, *Oper. Res.* **35** (1987) 291-305.

Gershwin, S. B., Representation and analysis of transfer lines with machines that have different processing rates, *Ann. Oper. Res.* **8** (1987) 511-530.

Gershwin, S. B., An efficient decomposition algorithm for unreliable tandem queueing systems with finite buffers, in: Perros and Altiok, (Eds.), *Proc First Int. Workshop on Queueing Networks with Blocking* (North-Holland, Amsterdam, 1989) 127-146.

Gershwin, S. B., Variance of output of a tandem production system, in: Onvural and Akyildiz, (Eds.), *Proc. Second International Workshop on Queueing Networks with Finite Capacity*, (North-Holland, Amsterdam, 1993) 291-304.

Gershwin, S. B. and Berman, O., Analysis of transfer lines consisting of two unreliable machines with random processing times and finite storage buffers, *AIIE Trans.* **13** (1981) 2-11.

Gershwin, S. B. and Shick, I. C., Modelling and analysis of three-stage transfer lines with unreliable machines and finite buffers, *Oper. Res* **31** (1983) 354-380.

Gordon, W. J. and Newell, G. F., Cyclic queueing systems with restricted length queues, *Oper. Res.* **15** (1967) 266-278.

Goto, K., Takahashi, Y., and Hasegawa, T., An approximate analysis of controlled tandem queues, in: *Proc. Conference on Modelling and Performance Evaluation*, (Springer, Berlin, 1983) 601-612.

Gün, L., *Annotated bibliography of blocking systems*, Tech. Rept., Electr. Eng. Dept., Univ. of Maryland, 1987.

Gün, L. and Makowski, A. M., Matrix-geometric solution for finite queues with phase-type distribution, in: Courtois and Latouche, (Eds.), *Performance '87* (North-Holland, 1987) 269-282.

Gün, L. and Makowski, A. M., An approximation method for general tandem queueing systems subject to blocking, in Perros and Altiok, (Eds.),*Proc. First International Workshop on Queueing Networks with Blocking*, (North-Holland, Amsterdam, 1989) 147-171.

Han, Y. and Smith, J. MacGregor, Approximate analysis of open M/M/C/K queueing networks, in: Onvural and Akyildiz, (Eds.), *Proc. Second International Workshop on Queueing Networks with Finite Capacity*, (North-Holland, Amsterdam, 1993) 113-126.

Hatcher, J. M., The effect of internal storage on the production rate of a series of stages having exponential service times, *AIIE Trans.* **1** (1969) 150-156. (See also: Letter to the above article by Knott, A. D., *AIIE Trans.* (1970) 273.)

Hendricks, K. B., The output processes of serial production lines of exponential machines with finite buffers, *Oper. Res.* **40** (1992) 1139-1147.

Hildebrand, D. K., Stability of finite queue, tandem server systems, *J. Appl. Probab.* **4** (1967) 571-583.

Hildebrand, D. K., On the capacity of tandem server, finite queue, service systems, *Oper. Res.* **16** (1968) 72-82.

Hillier, F. S. and Boling, R. W., Finite queues in series with exponential or Erlang service times - a numerical approach, *Oper. Res.* **15** (1967) 286-303.

Hillier, F. S. and So, K. C., The assignment of extra servers in tandem queueing system with small or no buffers, *Perform. Eval.* **10** (1989) 219-232.

Hong, Y., Glassey, C. R. and Seong, D., The analysis of a production line with unreliable machines and random processing times, *IIE Trans.* **24** (1992) 77-83.

Hordijk, A. and van Dijk, N., Networks of queues with blocking, in: Kylstra, (Ed.), *Performance '81* (North-Holland, Amsterdam, 1981) 51-65.

Hordijk, A. and van Dijk, N., Networks of queues: Part I: Job-local-balance and the adjoint process. Part II: General routing and service characteristics, in: Baccelli and Fayolle, (Eds.), *Modelling and Performance Evaluation Methodology*, (Springer-Verlag, Lecture Notes in Control and Information Science, Vol 60, 1983) 158-205.

Hsu, G.-H. and Yuan, X.-M., *Queueing system* $M^{(x)}/M/c \rightarrow /PH^{(r)}/1/K$, Tech. Rept., Institute of Applied Mathematics, Academia Sinica, 1991.

Hunt, G. C., Sequential arrays of waiting lines, *Oper. Res.* **4** (1956) 674-683.

Garofalakis, J. and Spirakis, P., *State space partition: A general approximation technique for solving non product form queueing systems*, Tech. Rept., Computer Science and Engineering Dept., University of Patras, 1989.

Ignall, E. and Silver, A., The output of a two-stage system with unreliable machines and limited storage, *AIIE Trans.* **9** (1977) 183-188.

Jafari, M. A. and Shanthikumar, J. G., *Allocation of buffer storages along a multi-stage automatic transfer line*, Tech. Rept. 84-003, Industrial Engineering Dept., Arizona University, 1984.

Jafari, M. A. and Shanthikumar, J. G., *A model of two-stage flow lines with non-geometric uptimes and downtimes and possible scrapping of workpieces*, Tech. Rept.85-008, Industrial Engineering and Operations Research Dept., Syracuse University, 1985.

Jafari, M. A. and Shanthikumar, J. G., *Determination of optimal buffer storage capacities and optimal allocation in multistage automatic transfer lines*, Tech. Rept. 85-011, Industrial Engineering and Operations Research Dept., Syracuse University, 1985

Jafari, M. A. and Shanthikumar, J. G., *Finite state spatially non-homogeneous quasi-birth-and-death-process*, Tech. Rept. 85-008, Industrial Engineering and Operations Research Dept., Syracuse University, 1985.

Jassim, A. R., *An M/G/1 queueing node with finite storage space and possibly blocked departure*, Tech. Rept., CSR, City University, 1987.

Jassim, A. R., *Performance modelling of finite storage double loop fault tolerant ring network*, Tech. Rept., CSR, City University, 1987.

Johri, P. K., Lipper, E. H., and Sengupta, B., Modeling and analysis of a production line with finite buffers and machines subject to breakdown, *RAIRO APII* **19** (1985) 471-483.

Jun, K.-P. and Perros, H. G., Approximate analysis of arbitrary configurations of queueing networks with blocking, in: Perros and Altiok, (Eds.), *Proc. First International Workshop on Queueing Networks with Blocking*, (North-Holland, Amsterdam, 1989) 259-280.

Jun, K.-P. and Perros, H. G., An approximate analysis of open tandem queueing networks with blocking and general service times, *Eur. J. Oper. Res.* **46** (1990) 123-135.

Kelly, F. P., *Reversibility and stochastic networks*, Wiley, 1979.

Kelly, F. P., The throughput of a series of buffers, *Adv. Appl. Prob.* **14** (1982) 633-653.

Kelly, F. P., Blocking, reordering, and the throughput of a series of servers, *Stoch. Process. Appl.* **17** (1984) 327-336.

Kerbache, L. and MacGregor Smith, J., The generalized expansion method for open finite queueing networks, *Eur. J. Oper. Res.* **32** (1987) 448-461.

Kingman, J. F. C., Markov population processes, *J. Appl. Prob.* **6** (1969) 1-18.

Knott, A. D., The inefficiency of a series of work stations - a simple formula, *Int. J. Prod. Res.* **8** (1970) 109-119.

Koenigsberg, E., Production lines and internal storage - a review, *Manage. Sci.* **5** (1959) 410-433.

Kogan, Y. A. and Krichagina, E. V., Closed exponential queueing networks with blocking in heavy traffic, in: Perros and Altiok, (Eds.), *Proc. First International Workshop on Queueing Networks with Blocking*, (North-Holland, Amsterdam, 1989) 217-226.

Kogan, Y. A. and Pukhalskii, A., On tandem queues with blocking in heavy traffic, in: Gelembe, (Ed.), *PERFORMANCE '84* (North-Holland, 1984) 549-558.

Konheim, A. G. and Reiser, M., A queueing model with finite waiting room and blocking, *J.ACM* **23** (1976) 328-341.

Konheim, A. G. and Reiser, M., Finite capacity queueing systems with applications in computer modelling, *SIAM J. Comput.* **7** (1978) 210-229.

de Koster, M. B. M., Capacity analysis of two-stage production lines with many products, *Engineering Costs and Production Economics* **12** (1987) 175-186.

de Koster, M. B. M., Estimation of line efficiency by aggregation, *Int. J. Prod. Res.* **25** (1987) 615-626.

de Koster, M. B. M., Approximation of flow lines with integrally controlled buffers, *IIE Trans.* **20** (1988) 374-381.

de Koster, M. B. M., Approximate analysis of production systems, *Eur. J. Oper. Res.* **37** (1988) 214-226.

de Koster, M. B. M., and Wijngaard, J., A continuous flow model for three production units in series with buffers, *Operations Research Proceedings 1985* (Springer, Berlin 1986) 253-264.

de Koster, M. B. M., and Wijngaard, J., On the use of continuous vs discrete models in production lines with buffers, in Perros and Altiok, (Eds.), *Proc. First Int. Workshop on Queueing Networks with Blocking* (North-Holland, Amsterdam, 1989) 175-191.

de Koster, M. B. M., and Wijngaard, J., Local and integral control of workload, *Int. J. Prod. Res.* **27** (1989) 43-52.

Kouvatsos, D. D. and Almond, J., Maximum entropy two-station cyclic queues with multiple general servers, *Acta Inf.* **26** (1988) 241-267.

Kouvatsos, D. D. and Denazis, S. G., *Comments on and tuning to: MEM for arbitrary queueing networks with repetitive-service blocking: A revision*, Tech. Rept. CS-18-91, Dept. of Computing, Univ. of Bradford, 1991.

Kouvatsos, D. D. and Denazis, S. G., *An efficient convolution procedure for general product-form approximations*, Tech. Rept. CS-18-92, Dept. of Computing, Univ. of Bradford, 1992.

Kouvatsos, D. D. and Denazis, S. G., Entropy maximized queueing networks with blocking and multiple job classes, *Perform. Eval.* **17** (1993) 189-205.

Kouvatsos, D. D. and Xenios, N. P., Maximum entropy analysis of general queueing networks with blocking, in: Perros and Altiok, (Eds.),*Proc. First International Workshop on Queueing Networks with Blocking*, (North-Holland, Amsterdam, 1989) 281-309.

Kouvatsos, D. D. and Xenios, N. P., MEM for arbitrary queueing networks with multiple general servers and repetitive-service blocking, *Perform. Eval.* **10** (1989) 169-195.

Kouvatsos, D. and Denazis, S. G., Entropy maximized queueing networks with blocking and multiple job classes, in: Onvural and Akyildiz, (Eds.), *Proc. Second International Workshop on Queueing Networks with Finite Capacity*, (North-Holland, Amsterdam, 1993) 127-142.

Krishna, C. and Shin, K., Queueing analysis of a canonical model of real-time multiprocessors, in: Bruell and Dowdy, (Eds.), *Proc. ACM SIGMETRICS Conference on Measurement and Modelling of Computer Systems* , (1983) 175-189.

Krisht, A. H. and Askin, R. G., *Approximate analysis of closed exponential queueing networks with finite buffers*, Tech. Rept., Systems and Industrial Engineering Dept., University of Arizona, 1988.

Krzesinski, A., *Approximate solution methods for multiclass queueing networks with state-dependent routing and blocking*, Tech. Rept., Univ. of Stellenbosch, 1987.

Kundu, S. and Akyildiz, I. F., Deadlock free buffer allocation in closed queueing networks, *Queueing Systems* **4** (1989) 47-56.

Labarta, J., Domingo, J., and Casals, O., Performance evaluation of packet switched omega networks with finite buffers, in: Perros and Altiok, (Eds.), *Proc. First International Workshop on Queueing Networks with Blocking*, (North-Holland, Amsterdam, 1989) 249-255.

Labetoulle, J. and Pujolle, G., A study of queueing networks with deterministic service and applications to computer networks, *Acta Inf.* **7** (1976) 183-195..

Labetoulle, J. and Pujolle, G., Modelling of packet switching communication networks with finite buffer size at each node, in: Chandy and Reiser, (Eds.), *Computer Performance*, (North-Holland, Amsterdam, 1977) 515-535.

Labetoulle, J. and Pujolle, G., Isolation method in a network of queues, *IEEE Trans. Soft. Eng.* **6** (1980) 373-381.

Langaris, C. and Conolly, B., On the waiting time of a two-stage queueing system with blocking, *J. Appl. Probab.* **21** (1984) 628-638.

Langaris, C. and Conolly, B., Two queues in tandem with multiple facilities and restricted waiting space, *Metron* **43** (1985).

Langaris, C. and Conolly, B., Three stage tandem queue with blocking, *Eur. J. Oper. Res* **19** (1985) 222-232.

Latouche, G. and Neuts, M. F., Efficient algorithmic solutions to exponential tandem queues with blocking, *SIAM J. of Algebraic Discrete Methods* **1** (1980) 93-106.

Lavenberg, S. S., Stability and maximum departure rate of certain open queueing networks having finite capacity constraints, *RAIRO Informatique/Computer Science* **12** (1978) 353-370.

Lazar, A. A. and Robertazzi, T. G., *The geometry of lattices for Markovian queueing networks*, Tech. Rept., Electrical Engineering Dept., Columbia Univ., 1984.

Lee, D.-S. and Sengupta, B., Two queues in tandem with dropping or blocking of customers, in: Onvural and Akyildiz, (Eds.), *Proc. Second International Workshop on Queueing Networks with Finite Capacity*, (North-Holland, Amsterdam, 1993) 1-13.

Lee, H.-S. and Pollock, S. M., Approximate analysis for the merge configuration of an open queueing network with blocking, *IIE Trans.*, **21** (1989) 122-129.

Lee, H.-S. and Pollock, S. M., Approximate analysis of open acyclic exponential queueing networks with blocking, *Oper. Res.* **38** (1990) 1123-1134.

Lee, W. T., Tcha, D.-W., and Yamazaki, G., *Server assignment for multi-stage production systems with finite buffers*, Tech. Rept., Management Sci. Dept, KAIST, 1990.

Le Ny, L.-M., Étude analytique de reseaux de files d'attende multiclasses à routage variable, *RAIRO Recherche Operationnelle /Oper Res* **14** (1980) 331-347.

Le Ny, L.-M., Forme produit pour des reseaux multiclasses à routage dynamique, *Annales Scientifiques de l'Universitè de Clermont-Ferrand II* **76** (1983) 17-34.

Liebeherr, J., Deadlock-free buffer allocations in multiple chain blocking networks with tandem sequences, in: Onvural and Akyildiz, (Eds.), *Proc. Second International Workshop on Queueing Networks with Finite Capacity*, (North-Holland, Amsterdam, 1993) 41-54.

Liebeherr, J. and Akyildiz, I. F., *Deadlock freedom in queueing networks with multiple routing chains and finite capacities*, Tech. Rept., College of Computing, Georgia Institute of Technology, 1991.

Liu, X.-G. and Buzacott, J. A., A balanced local flow technique for queueing networks with blocking, in: Perros and Altiok, (Eds.), *Proc. First International Workshop on Queueing Networks with Blocking*, (North-Holland, Amsterdam, 1989) 87-104.

Liu, X.-G., Zhuang, L., and Buzacott, J. A., A decomposition method for throughput analysis of cyclic queues with production blocking, in: Onvural and Akyildiz, (Eds.), *Proc. Second International Workshop on Queueing Networks with Finite Capacity*, (North-Holland, Amsterdam, 1993) 253-266.

Liu, Z. and Resing, J. A. C., *Duality and equivalencies in closed tandem queueing networks*, Tech. Rept. COSOR 93-30, Math. and Comp. Sci. Dept., Eindhoven Univ., 1993.

MacGregor Smith, J. and Daskalaki, S., Buffer space allocation in automated assembly lines, *Oper. Res.* **36** (1988) 343-358.

Makino, T., On the mean passage time concerning some queueing problems of the tandem type, *J. Oper. Res. Soc. Japan* **7** (1964) 17-47.

di Mascolo, M., David, R., and Dallery, Y., *Modelling and analysis of assembly systems with unreliable machines and finite buffers*, Tech. Rept. 88-99, Laboratoire d'Automatic de Grenoble, 1988.

Massey, W. A., Balanced queues in series with communication blocking, *Math. Oper. Res.*, to appear, 1993.

McClain, J. O. and Moodie, D. R., A comment on "buffer space allocation in automated assembly lines", *Oper. Res.* **39** (1991) 857-860.

McCormick, S. T., Pinedo, M., Shenker, S., and Wolf, B., Sequencing in an assembly line with blocking to minimize cycle time, *Oper. Res.* **37** (1989) 925-935.

Merlin, P. M. and Schweitzer, P. J., Deadlock avoidance in store-and-forward networks-I: Store-and forward deadlock, *IEEE Trans. Comm.* **28** (1980) 345-354.

Melamed, B., A note on the reversibility and duality of some tandem blocking queueing systems, *Manage. Sci.* **32** (1986) 1648-1650.

Mitra, D. and Mitrani, I., Analysis of a kanban discipline for cell coordination in production lines I, *Manage. Sci.* **36** (1990) 1548-1566.

Mitra, D. and Mitrani, I., Analysis of a kanban discipline for cell coordination in production lines II: Stochastic demands, *Oper. Res.* **39** (1991) 807-823.

Morris, T. D. and Perros, H. G., Performance modelling of a multi-buffered Banyan switch under bursty traffic, *INFOCOM '92*, Florence, Italy, May 1992, 436-445.

Morris, T. D. and Perros, H. G., Approximate analysis of a discrete-time tandem network of cut-through queues with blocking and bursty traffic, in: Onvural and Akyildiz, (Eds.), *Proc. Second International Workshop on Queueing Networks with Finite Capacity*, (North-Holland, Amsterdam, 1993) 55-68.

Muth, E. J., The production rate of a series of work stations with variable service times, *Int. J. Prod. Res.* **11** (1973) 155-169.

Muth, E. J., The reversibility property of production lines, *Manage. Sci.* **25** (1979) 152-158.

Muth, E. J. and Yeralan, S., Effect of buffer size on productivity of work stations that are subject to breakdowns, in: *Proc. 20th IEEE Conf. on Decision and Control*, (1981), 643-648.

Neuts, M. F., Two-queues in series with a finite, intermediate waiting room, *J. Appl. Probab.* **5** (1968) 123-142.

Neuts, M. F., *Matrix-geometric solutions in stochastic models - an algorithmic approach*, (The John Hopkins Univ. Press, 1981) 232-245.

Newell, G. F., *Approximate behavior of tandem queues*, Lecture Notes in Economics and Mathematical Systems, **171** (Springer, Berlin, 1979).

Nilsson, A. A., and Altiok, T., Open queueing networks with finite capacity queues, in: *Proc. 1981 Int. Conf. Parallel Processing,* (1981), 87-91.

De Nitto Personè, V., *Topology related index for performance comparison of blocking symmetrical networks*, Tech. Rept., Dipartmento di Ingegneria Elettronica, Universita di Roma, 1990.

De Nitto Personè, V. and Grillo, D., *Impact of blocking policies in closed, exponential and symmetrical networks*, Tech. Rept. #4B0586, Networks and Services Dept., Fondazione Ugo Bordoni, Italy, 1986.

De Nitto Personè, V. and Grillo, D., Managing blocking in finite capacity symmetrical ring networks, in: Moraes, de Souza e Silva, and Soares, (Eds.), *Proc. of 3rd Int. Conf. on Data Communication Systems and their Performance*, Rio de Janeiro, 1987, 225-240.

Ohmi, T., An approximation for the production efficiency of automated transfer lines with in-process storage, *AIIE Trans.* **13** (1981) 22-28.

Okamura, K. and Yamashina, H., Analysis of the effect of buffer storage capacity in transfer line systems, *AIIE Trans.* **9** (1977) 127-135.

Onvural, R. O., *Closed queueing networks with finite buffers*, Ph.D. Thesis, North Carolina State University, 1987.

Onvural, R. O., On the exact decomposition of closed queueing networks with finite buffers, in: Perros and Altiok, (Eds.), *Proc. First International Workshop on Queueing Networks with Blocking*, (North-Holland, Amsterdam, 1989) 73-83.

Onvural, R. O., A note on the product-form solutions of multiclass closed queueing networks with blocking, *Perform. Eval.* **10** (1989) 247-253.

Onvural, R. O., A survey of closed queueing networks with finite buffers, *ACM Computing Surveys*, **22** (1990) 83-121. (A version of this paper also appeared in: Takagi (Ed.) *Stochastic Analysis of Computer and Communication Systems* (North-Holland, Amsterdam).

Onvural, R. O. and Perros, H. G., On equivalencies of blocking mechanisms in queueing networks with blocking, *Oper. Res. Lett.* **5** (1986) 293-297.

Onvural, R. O. and Perros, H. G., Some equivalencies between closed queueing networks with blocking, *Perform. Eval.* **9** (1988/89) 111-118.

Onvural, R. O. and Perros, H. G., Equivalencies between open and closed queueing networks with finite buffers, *Perform. Eval.* **9** (1988/89) 263-269.

Onvural, R. O. and Perros, H. G., Throughput analysis of cyclic queueing networks with finite buffers, *IEEE Trans. Soft. Eng.* **15** (1989) 800-808.

Onvural, R. O. and Perros, H. G., Approximate analysis of multi-class tandem open queueing networks with Coxian parameters and finite buffers, in: King, Mitrani, and Pooley, (Eds.), *Performance '90*, (North-Holland, 1990) 131-140.

Onvural, R. O., Perros, H. G., and Altiok, T., On the complexity of the matrix-geometric solution of exponential open queueing networks with blocking, in: Pujolle, Fdida, and Horlait, (Eds.), *Proc. Int. Workshop on Modelling Techniques and Performance Evaluation*, (North-Holland, Amsterdam, 1987) 3-12.

Paik, C. and Tcha, D.-W., *Throughput equivalencies in fork/join queueing networks with finite buffers and general service times*, Tech. Rept., Management Sci. Dept., KAIST, Korea, 1993.

Papadopoulos, H. T. and O'Kelly, M. E. J., *Exact analysis of production lines,* Tech. Rept., Industrial Engineering Dept., University College, Galway, 1988.

Patterson, R. L., Markov processes occurring in the theory of traffic flow through an N-stage stochastic service system, *J. Ind. Eng.* **14** (1964) 188-193.

Pellaumail, J. and Boyer, P., *Deux files d'attente à capacité limitée en tandem*, Tech. Rept. 147, CNRS/INRIA - Rennes, 1981.

Perros, H. G., A symmetrical exponential open queue network with blocking and feedback, *IEEE Trans. Soft. Eng.* **7** (1981) 395-402.

Perros, H. G., A two-level open queue network with blocking and feedback, *RAIRO Recherche Operationnelle/Oper. Res.* **15** (1981) 27-38.

Perros, H. G., *A survey of two-node queueing networks with blocking*, Tech. Rept., Computer Science Dept., North Carolina State University, 1988.

Perros, H. G., A bibliography of papers on queueing networks with with finite capacity queues, *Perform. Eval. Review* **10** (1989) 225-260.

Perros, H. G., Approximation algorithms for open queueing networks with blocking, in Takagi (Ed.) *Stochastic Analysis of Computer and Communication Systems*, (North-Holland, 1990) 451-498.

Perros, H. G. and Altiok, T., Approximate analysis of open networks of queues with blocking: tandem configurations, *IEEE Trans. Soft. Eng.* **12** (1986) 450-461.

Perros, H. G., Nilsson, A. A. and Liu, Y.-C., Approximate analysis of product-form type queueing networks with blocking and deadlock, *Perform. Eval.* **8** (1988) 19-39.

Perros, H. G. and Snyder, P. M., A computationally efficient approximation algorithm for analyzing queueing networks with blocking, *Perform. Eval.* **9** (1988/89) 217-224.

Pinedo, M. and Wolf, R. W., A comparison between tandem queues with dependent and independent service times, *Oper. Res.* **30** (1982) 464-479.

Pittel, B., Closed exponential networks of queues with saturation: The Jackson-type stationary distribution and its asymptotic analysis, *Math. Oper. Res.* **4** (1979) 367-378.

Pollock, S. M., Birge, J. R., and Alden, J. M., *Approximation analysis for open tandem queues with blocking: exponential and general distributions,* Tech. Rept., 85-30, IOE Dept., Univ. of Michigan, 1985.

Prabhu, N. U., Transient behavior of a tandem queue, *Manage. Sci.* **13** (1967) 631-639.

Pujolle, G. and Potier, D., Reseaux de files d'attente à capacité limitée avec des applications aux systemes informatiques, *RAIRO Informatique/Computer Science* **13** (1979) 175-197.

Rao, N. P., On the mean production rate of a two-stage production system of the tandem type, *Int. J. Prod. Res.* **13** (1975) 207-217.

Rao, N. P., Two-stage production systems with intermediate storage, *AIIE Trans.* **7** (1975) 414-421.

Rege, K., Approximate analysis of serial manufacturing lines with buffer control, *Large Scale Systems J*, to appear, 1989.

Sakasegawa, H. and Kawashima, T., *On the equivalence of three types of blocking in non-Markovian tandem queues*, Tech. Rept., Institute of Socio-Economic Planning, Univ. of Tsukuba, 1991.

Schweitzer, P. J. and Altiok, T., Aggregate modelling of tandem queues with intermediate buffers, in: Perros and Altiok, (Eds.), *Proc. First International Workshop on Queueing Networks with Blocking*, (North-Holland, Amsterdam, 1989) 19-29.

Sevast'yanov, B. A., Influence of storage bin capacity on the average standstill time of a production line, *Teoriya Veroyatnostey i ee Primeneniya* **7** (1962) 429-438 (English translation: *Theory of Probability and its Applications* **7** (1962) 429-438).

Shantikumar, J. G. and Jafari, M. A., *Bounding the performance of tandem queues with finite buffer space*, Tech. Rept., School of Bus. Amin., Univ. of California at Berkeley, 1987.

Shanthikumar, J. G. and Yao, D. D., Monotonicity properties in cyclic queueing networks with finite buffers, in: Perros and Altiok, (Eds.), *Proc. First International Workshop on Queueing Networks with Blocking*, (North-Holland, Amsterdam, 1989) 325-344.

Sheskin, T. J., Allocation of interstage storage along an automatic production line, *AIIE Trans.* **8** (1976) 146-152.

Shiue, G. A. and Altiok, T., Two-stage, multi-product production/inventory systems, in: Onvural and Akyildiz, (Eds.), *Proc. Second International Workshop on Queueing Networks with Finite Capacity*, (North-Holland, Amsterdam, 1993) 213-223.

So, K. C. and Chin, K. E., *Performance bounds on multiserver exponential tandem queues with finite buffers*, Tech. Rept.

DS90007, Graduate School of Management, University of California, Irvine, 1990.

Soyster, A. I. and Toof, D. I., Some comparative and design aspects of fixed cycle production systems, *Naval Res. Logist. Quart.* **23** (1976) 437-454.

Soyster, A. L., Schmidt, J. W., and Rohrer, M. W., Allocation of buffer capacities for a class of fixed cycle production lines, *AIIE Trans.* **11** (1979) 140-146.

Stewart, W. J., Recursive procedures for the numerical solution of Markov chains, in: Perros and Altiok, (Eds.), *Proc. First International Workshop on Queueing Networks with Blocking*, (North-Holland, 1989) 229-247.

Suri, R. and Diehl, G. W., A new building block for performance evaluation of queueing networks with finite buffers, *Proc. ACM SIGMETRICS on Measurement and Modelling of Computer Systems*, (1984) 134-142.

Suri, R. and Diehl, G. W., A variable buffer-size model and its use in analyzing closed queueing networks with blocking, *Manage. Sci.* **32** (1986) 206-224.

Suzuki, T., On a tandem queue with blocking, *J. Oper. Res. Soc. Japan* **6** (1964) 137-157.

Takahashi, Y., Aggregate approximation for acyclic queueing networks with communication blocking, in: Perros and Altiok, (Eds.), *Proc. First International Workshop on Queueing Networks with Blocking*, (North-Holland, Amsterdam, 1989) 33-46.

Takahashi, Y., Miyahara, H. and Hasegawa, T., An approximation method for open restricted queueing networks, *Oper. Res.* **28** (1980) 594-602.

Tsiotras, G. D., Modeling and analysis of multiple-class tandem queueing systems with finite capacities, *Math. Comput. Modeling*,1990.

Tsiotras, G. D. and Badr, H., Modeling the blocking phenomenon of service facilities, *Math. Comput. Modeling* **12** (1989) 641-649.

Tsiotras, G. D. and Badr, H., A recursive methodology for the derivation of the blocking probabilities of tandem queues with finite capacity. *Computers and Operations Research* **17** (1990) 475-479.

Tsiotras, G. D., Badr, H., and Beltrami, E., Blocking probabilities for a class of two-station tandem queueing models, *IIE Trans.* **23** (1992) 331-345.

Tsoucas, P. and Walrand, J., Monotonicity of throughput in non-Markovian networks, *J. Appl. Probab.* **26** (1989) 134-141.

van Dijk, N. M., A simple bounding methodology for non-product-form finite capacity queueing networks, in: Perros and Altiok, (Eds.), *Proc. First International Workshop on Queueing Networks with Blocking*, (North-Holland, Amsterdam, 1989) 3-17.

van Dijk, N. M., Comment on Yao and Buzacott's "modeling a class of flexible manufacturing systems with reversible routing", *Oper. Res.* **37** (1989) 845-846.

van Dijk, N. M., A simple throughput bound for large closed queueing networks with finite capacities, *Perform. Eval.* **10** (1989) 153-167.

van Dijk, N. M., On 'stop=repeat' servicing for non-exponential queueing networks with blocking, *J. Appl. Probab.* **28** (1991a) 159-173.

van Dijk, N. M., 'Stop=recirculate' for exponential product-form queueing networks with departure blocking, *Oper. Res. Lett.* **10** (1991b) 343-351.

van Dijk, N., A note on the arrival theorem for closed queueing networks with blocking, in: Onvural and Akyildiz, (Eds.), *Proc. Second International Workshop on Queueing Networks with Finite Capacity*, (North-Holland, Amsterdam, 1993) 281-290.

van Dijk, N. and Lamond, B. F., Bounds for the call congestion in tandem queues, *Oper. Res.* **36** (1988) 470-477.

van Dijk, N. M. and Tijms, H. G., Insensitivity in two node blocking models with applications, in: Boxma and Cohen, (Eds.), *Teletraffic Analysis and Computer Performance Evaluation*, (North-Holland, Amsterdam, 1986) de Waal, P. and van Dijk, N. M., *Interconnected networks of queues with randomized arrival and departure blocking*, Tech. Rept., Econometrics Dept., Free University, 1989.

Wesfreid, E., *Dynamic behaviour of N-stage transfer lines with unreliable machines and finite buffers*, Tech. Rept. 456, INRIA , 1985.

Wijngaard, J., The effect of interstage buffer storage on the output of two unreliable production units in series with different production rates, *AIIE Trans.* **11** (1979) 42-47.

Wong, B., Giffin, W. and Disney, R., Two-finite M/M/1 queues in tandem: a matrix solution for the steady state, *Opsearch* **14** (1977) 1-18.

Xenios, N. P., *General queueing networks with blocking*, Ph.D. thesis, University of Bradford, 1989.

Xie, X., *Performance analysis of a transfer line with unreliable machines and finite buffers*, Tech. Rept. 1089, INRIA, 1989.

Yamashita, H., *Approximate aggregation for arbitrary configurations of queueing networks with shared buffer and blocking*, Tech. Rept., Mechanical Engineering Dept., Sophia University, 1990.

Yamashita, H., Nakazawa, K., Nagasuka, H., and Suzuki, S., Modeling and analysis of a production line operated according to demand, *JSME* **32** (1989) 491-1497.

Yamashita, H. and Suzuki, S., An approximation method for line production rate of a serial production line with common buffer, *Comp. and Oper. Res.* **15** (1988) 395-402.

Yamazaki, G., Kawashima, T., and Sakasegawa, H., Reversibility of tandem blocking queueing systems, *Manage. Sci.* **31** (1985) 78-83.

Yao, D. D. and Buzacott, J. A., Modeling a class of state-dependent routing in flexible manufacturing systems, *Ann. Oper. Res.* **3** (1985) 153-167.

Yao, D. D. and Buzacott, J. A., The exponentialization approach to flexible manufacturing system models with general service times, *Eur. J. Oper. Res.* **24** (1986) 410-416.

Yao, D. D., and Buzacott, J. A., Modeling a class of flexible manufacturing systems with reversible routing, *Oper. Res.* **35** (1987) 87-93.

Ziedins, I., *Tandem queues with correlated service times and finite capacity*, Tech. Rept., Dept. of Actuarial Mathematics and Statistics, Heriot-Watt University, 1989.

Table of Symbols

The main symbols that have been used in the book and their explanation are listed below. The symbols are given per Chapter, and in the order in which they appear within each section. Some commonly used symbols are given separately.

Commonly Used Symbols

M	Number of nodes in a queueing network with blocking
r_{ij}	Branching probabilities in a queueing network with blocking
m_i	Buffer capacity of node i
λ	Rate of arrivals
μ_i	Service rate at the ith node
ρ	Traffic intensity
a_{ij}	Branching probabilities in a phase-type distribution
$(\alpha, \mathbf{T})$	Representation of a phase-type distribution
n_i	Random variable indicating the number of customers in node i
$(n_1,\ldots,n_M)$	M-tuplet indicating the state of a queueing network with blocking
$p(n_1,\ldots,n_M)$	Joint queue-length probability distribution of a queueing network with blocking
$p_i(n_i)$	Probability that there are n_i customers in node i
N	Number of customers in a closed queueing network with blocking
ε	A small number used to test convergence of an algorithm

Chapter 1

Section 1.4

C_k	Coxian distribution with k phases

PH_k	Phase-type distribution with k phases
H_k	Hyper-exponential distribution with k phases
GE	Generalized exponential distribution
μ_i	Service rate of the ith phase in a phase-type distribution
a_i	Branching probability in a Coxian distribution
p_i	Branching probability in a hyper-exponential distribution
GE(r,ra)	Representation of a GE distribution
f(x)	pdf of a random variable X
F(x)	Cumulative distribution of a random variable X
f*(s)	Laplace transform of f(x)
c^2	Squared coefficient of variation
m_i	ith moment of a random variable

Chapter 2

Section 2.2

$p_{ij}(s,t)$	Probability that the process is in state j at time t given that it was in state i at time s
$\mathbf{P}(t)$	Transition probability matrix $(p_{ij}(t,t+\Delta t))$
$q_{ij}(t)$	Rate of transition from state i to state j at time t
$\mathbf{Q}(t)$	Rate matrix $(q_{ij}(t))$
$x_i(t)$	Probability that the process is in state i at time t
$\mathbf{x}(t)$	Probability vector $(x_1(t),x_2(t),...,x_n(t))$
$\mathbf{x}$	Probability vector $(x_1,...,x_n)$

Section 2.4

$\mathbf{y}$	Transposed $\mathbf{x}$, i.e. $\mathbf{y}=\mathbf{x}^T$
$\mathbf{L}, \mathbf{U}$	LU decomposition matrices of $\mathbf{Q}$, i.e. $\mathbf{Q}=\mathbf{LU}$
$\mathbf{e}_i$	Column vector of 0's where the ith row is a 1
$\mathbf{B}, \mathbf{D}, \mathbf{H}$	Decomposition matrices of $\mathbf{Q}^T$, i.e. $\mathbf{Q}^T=\mathbf{B}+\mathbf{D}+\mathbf{H}$

Section 2.5

$\mathbf{R}$	Matrix $\mathbf{R}$ in the matrix-geometric procedure
$\mathbf{e}$	Column vector of 1's

Chapter 3

Section 3.2

m	Buffer capacity of the second node
$g_k(z)$	Generating function
μ_1^0	Service rate of the first server below which the first node gets saturated
σ	$=\mu_1/\mu_2$
σ_0	$=\mu_1^0/\mu_2$
π^∞, π^0	Blocking probability for the case when $\mu_1 \to \infty$ and when $\mu_1 \leq \mu_1^0$
k_i	Number of servers in the ith node
s_i	State of the ith server

Section 3.3

p_i, r_i	Breakdown and repair probabilities for server i
$\bar{p}_i$	$=1-p_i$
$\bar{r}_i$	$=1-r_i$

Section 3.4

$(\lambda_1, \lambda_2, a_1)$	Parameters of the Coxian service time of the first server
(μ_1, μ_2, a_2)	Parameters of the Coxian service time of the second server
$\mathbf{p}(n)$	Probability sub-vector that there are n customers in node 2
$\otimes$	Kronecker product of two matrices

Chapter 4

Section 4.2

T	Throughput
$\bar{\mu}_i$	Effective service rate at node i

$\bar{\lambda}_i$	Effective arrival rate to node i
λ_i	Overall arrival rate to node i (when node i is treated in isolation)
π_i	Conditional probability that upon a service completion at node i, node i+1 is full
$w_i(j)$	Probability that the ith server is in the jth phase of its service time, given that the ith node is full
(n_i,s_i)	State vector of the ith node; n_i indicates the number of customers in node i and s_i the state of the ith server

Section 4.3

$D_i^*(s)$	Laplace transform of the inter-departure time from node i
$B_i^*(s)$	Laplace transform of the effective service time at node i
$E_i^*(s)$	Laplace transform of the effective inter-arrival time at node i
$O_i^*(s)$	Laplace transform of the inter-arrival time of the overall arrival process at node i (when the node is treated in isolation)
f_i	Probability that an arriving customer finds node i full
$p_i^+(n_i)$	Probability that node i contains n_i customers immediately after a departure
$m_{ei,j}$	jth moment of the effective arrival process at node i
$m_{oi,j}$	jth moment of the overall arrival process at node i (when the node is treated in isolation)
δ_i	Probability that a departing customer leaves node i empty
$\beta_i(j)$	Probability that the upstream server of sub-system i is in its jth phase at the time when the downstream server becomes idle

Section 4.4

Y_i	Constant service time at node i

k^*	Node with the longest service time
A_n	Arrival time at the queueing network of the nth customer
D_n	Departure time from the queueing network of the nth customer
S_n	Total time the nth customer spent in the queueing network
W_n	Total time the nth customer spent queueing in the queueing network
d_i	Difference between the total amount of service already received by the (n-1)st customer and the nth customer at the point of departure of the (n-1)st customer from node i
$W_n(i)$	Delay experienced by the nth customer in the ith node
$W(\bar{i})$	Average waiting time of a customer in node i
$\mathcal{W}_n$	Total time the nth customer spent in the queueing network
$\bar{\mathcal{W}}$	Average time a customer spent in the queueing network
T_L, T_U	Lower and upper bound of the throughput of a tandem configuration

Section 4.5

p_i, r_i	Breakdown and repair probabilities of server i
$p_u(i)$, $r_u(i)$	Breakdown and repair probabilities for the upstream server of the ith sub-system
$p_d(i)$, $r_d(i)$	Breakdown and repair probabilities for the downstream server of the ith sub-system.
p_{uu}	Conditional probability that the upstream server of the (i-1)st subsystem is down given that the upstream server of the ith subsystem is down
p_{dd}	Same as above, but for downstream servers
$T(i)$	Throughput of the ith sub-system
$p_0(i)$, $p_f(i)$	Probability that the ith subsystem is empty, full
o_i	% of time that server i is operational

Chapter 5

Section 5.2

λ_1	External rate of arrivals to node 1
$\bar{\lambda}_i$	Effective arrival rate to node i
λ_i	Overall arrival rate to node i (when node i is treated in isolation)
$\pi_{ij}(n_j)$	Conditional probability that upon service completion at server i, there are n_j customers in node j
$\pi_j(n_j)$	Probability that an arriving customer to node j will see n_j customers
ν_{ijk}	Service rate of a phase in a phase-type distribution

Section 5.3

$H(p)$	Entropy function
$<f_i(n)>$	Expectation
$f_i(n)$	Function of system state n
G	Normalizing constant
θ_i	Lagrange multiplier
$(s,s\lambda)$	GE inter-arrival process
$(r,r\mu)$	GE service time process
$\zeta_0(n_i)$	$=\min\{1, \max(0,n_i)\}$
$\zeta_i(n_i)$	$=\max\{0, n_i-m_i+1\}$
L	Mean queue length in a node
f	Probability a node is full
γ_{ij}	jth Langrangian multiplier for node i
(r_i,r_ia_i)	GE parameters of node i's service time
μ_i, c_{si}^2	Service rate and squared coefficient of variation of the service time at node i
π_{ij}	Probability that a customer upon completion of its service at node i is blocked by node j
π_{di}	Probability that a customer is blocked upon completion of its service at node i

$\bar{\mu}_i$, $\bar{c}^2_{si}$	Service rate and squared coefficient of variation of the effective service time at node i
λ_i, c^2_i	Arrival rate and squared coefficient of variation of the inter-arrival time of the overall arrival process to node i (when treated in isolation)
$\bar{\lambda}_i$, $\bar{c}^2_i$	Arrival rate and squared coefficient of variation of the inter-arrival time of the effective arrival process to node i
λ_{0i}, c^2_{0i}	Arrival rate and squared coefficient of variation of the inter-arrival time of the external arrival process to node i
π_{0i}	Probability that an externally arrived customer finds node i full
$\bar{c}^2_{di}$	Squared coefficient of variation of the inter-departure time from node i
$\bar{c}^2_{dji}$	Squared coefficient of variation of the effective arrival process from node j to node i

Chapter 6

Section 6.2

X(t)	Stochastic process
q(n,n′)	Transition rate from state n to state n′
Y(t)	Truncated process of X(t)
p(n)	Probability that the system is in state n
$p_A(n)$	Probability that the truncated process is in state n

Section 6.3

n_0, n_u	Minimum and maximum occupancy in node 1 in a two-node closed queueing network with blocking

Section 6.4

C	Number of classes of customers in a closed queueing network with blocking
$r_{ic,jd}$	Branching probabilities in a multi-class queueing network with blocking

$f_i(n_i)$	Service capacity of the server in the ith node
$b_j(n_j)$	Probability that a customer will be admitted to node j, when there are n_j customers in it
e_i	Parameters satisfying traffic equations: $e_i r_{ij}=e_j r_{ji}$
$\mathbf{1}_i$	Row vector of zeroes with the ith column containing a 1
G	Normalizing constant
F_{ic}	Service distribution for class c in node i
μ_{ic}	Service rate at node i for class c
$g_{ic,p}$	Probability that a class c customer in node i requests an Erlang distributed service with p phases
$E_{p\nu_{ic}}$	Erlang distribution with p phases, and a node and class dependent phase service time ν_{ic}
$R_{ic}(p)$	Probability that at an arbitrary instant, a class c customer in service at node i has p phases more to traverse
n_{ic}	Number of class c customers present in the ith node
c_{il}	Class of customer occupying the lth position in the ith node
$\gamma_i(l,n_i)$	Fraction of the service capacity destined to the customer in position l when there are n_i customers in node i
$\xi_i(l,n_i)$	Probability that an arriving customer to node i is placed in position l when there are n_i customers in node i
$g(z)$	Generating function
$\tilde{g}_i(z)$	Partial product
$G(i,j)$	Coefficient of the jth term of $\tilde{g}_i(z)$

Section 6.5

$\zeta_i(n_i)$, $\eta_i(n_i)$	Functions of n_i, the number of customers in node i
b	Indicates the node with the smallest buffer capacity

Chapter 7

Section 7.2

ℓ	Indicates the node with the largest buffer capacity
b	Indicates the node with the smallest buffer capacity
B	Total buffer capacity in a closed queueing network with blocking
$p^N(.)$	Queue-length probability distribution of a closed queueing network with blocking with N customers
$L_i(N)$	Mean number of customers in node i when there are N customers in the closed queueing network with blocking
L_i^{QNj}	Mean number of customers in the ith node of the closed queueing network with blocking QNj

Section 7.3

T(N)	Throughput of the cyclic queueing network when there are N customers in it
$\mathbf{k}=(k_1,...,k_M)$	Initial distribution of customers in a cyclic queueing network with blocking
$\mathbf{h}=(h_1,...,h_M)$	Initial distribution of holes in a cyclic queueing network with blocking
$t_{i,n}$	Service time of the nth customer at the ith node
$C_{i,n}$	Time at which the nth customer begins its service at the ith node
$D_{i,n}$	Time at which the nth customer completes its service at the ith node
D_n	Time at which n service completions have occurred at each server
s, r, d	Superscripts referring to the symmetrical, reverse, and dual cyclic queueing network respectively

Section 7.4

N^*	Number of customers for which the throughput is maximum
α, α', χ	Unknown parameters

Section 7.5

Ω_i	ith lump
ω_i	number of states in ith lump
ω	Number of lumps

Section 7.6

$\lambda_i(n_i)$, $\mu_i(n_i)$	State-dependent arrival and service rate for node i
$\tilde{n}_i$	Smallest number of customers always present in node i
u_i	Probability that there are more than $\tilde{n}_i$ customers in node i
$\upsilon_i(n_i)$	$=\max\{0,n_i-(\tilde{n}_i+1)\}$
$\zeta_i(n_i)$	$=\max\{0,n_i-m_i+1\}$
f_i	Probability node i is full
$\tilde{x}_i$, $\tilde{y}_i$, $\tilde{z}_i$	Lagrange coefficients
$\bar{\lambda}_{0i}$	Effective external arrival rate
L_i	Mean number of customers in node i of the pseudo open queueing network with blocking

(See also section 5.3)

Chapter 8

Section 8.2

(R,r)	Inventory control policy
d_i	Probability an order requires i finished products
Y_i, X_i	Service time and set-up time at the ith server

Section 8.3

$\bar{t}_b$	Mean time a blocking customer spends in the second and third-level servers
J_1, J_2	Conditional expectation of the number of blocked first-level servers

Section 8.4

r_{uu}, r_{ul}, r_{lu}, r_{ll}	Branching probabilities in a switching element
p, q, λ	Parameters of an Interrupted Bernoulli Process (IBP)
π	Probability that a cell at the head of an output buffer is blocked at the beginning of a switch cycle due to the destination input buffer being full
P	Probability transition matrix
x	Stationary probability vector
A	Set of active states
I	Set of idle states
$q(\mathbf{n},\mathbf{n}')$	Transition probability from state **n** to state **n′**
p_u, q_u, λ_u	Parameters of the arrival process to the upper input buffer of a switching element
p_l, q_l, λ_l	Parameters of the arrival process to the lower input buffer of a switching element

Index